STORMY WEATHER

Stormy Weather

PAGAN COSMOLOGIES, CHRISTIAN TIMES, CLIMATE WRECKAGE

William E. Connolly

FORDHAM UNIVERSITY PRESS NEW YORK 2024

Fordham University Press also publishes its books in a variety of electronic formats. Some content that appears in print may not be available in electronic books.

Visit us online at www.fordhampress.com.

Library of Congress Cataloging-in-Publication Data available online at https://catalog.loc.gov.

Printed in the United States of America

26 25 24 5 4 3 2 1

First edition

Contents

To uproot depraved and ancient opinions.

—AUGUSTINE

The limitless expanse of the sea echoed terribly; the earth rumbled loudly, and the broad reach of the sky shook and groaned.

—HESIOD

The Indians occupied but did not possess the land. It is by agriculture that man wins the soil.

—TOCQUEVILLE

The West does not exist. I know, I've been there.

—MICHEL-ROLPH TROUILLOT

By feeling the becoming of the other, the divine itself becomes.

—CATHERINE KELLER

Is "a politics of vitality" at all possible, or desirable, at this advanced stage of the planetary crisis?

—AMITAV GHOSH

The way humans perceive animals and other subjectivities that inhabit the world . . . differs profoundly from how they see themselves.

—EDUARDO VIVEIROS DE CASTRO

[To Aztecs] human life is consequently unstable, fragile, perilous, fleeting and evanescent.

—JAMES MAFFIE

The rapidity with which the climate switched from icehouse to hothouse conditions was particularly astounding, with hippos . . . from the tropics roaming England and other northern temperature zones just a few thousand years after the landscape was buried beneath ice or frozen solid.

—BILL MCGUIRE

Life is tragic, simply because the earth turns and the sun inexorably rises and sets.

—JAMES BALDWIN

I am trying to think time as a multiplicity.

—MICHEL SERRES

Introduction

Lived Cosmologies and Climate Wreckage

1.

A lived cosmology, to me, consists of embedded understandings about the beginning of the cosmos and earth, the trajectories time then follows, and unfolding relations between events and gods, god, or godless processes. A cosmology, oriented either to the priority of god(s) (and eminent causality) or to ungodly forces (and emergent causality), informs how people live. It often slumbers in the background until disturbed by an unexpected planetary event, economic change, invasion, religious movement. or scientific shift. A disturbing event may intensify faith in the projections of a received cosmology or it may prompt adjustments within it. It depends. Lived cosmologies of the west have played roles of significance in the routines people follow, the faiths they confess, the economies they fashion, their orientations to other humans and species, and their operative dispositions to the earth.[1]

Take, for example, recent struggles in the American west over the rapidly depleting Ogallala aquifer that has sustained expanding settler populations there for over two hundred years. Many ranchers, real estate investors, banks, and oil interests fight to continue the depletion even though it is demonstrably unsustainable. Indigenous peoples and ecologists fight valiantly against them. What about those in between? Where do they go? One statement by the president of the Kansas Farm Bureau is revealing. If you cut water use, he says, "you lose land value, you lose a tax base, and quite frankly you lose an entire way of life."[2] An "entire way of life," of course, includes the right to ravage the land and drain the aquifer. It also construes a multifaceted image of temporal progress lurking inside that statement to be sacrosanct, even if it now holds on by its fingertips. So promoters pretend things won't become disastrous, and others allow them to do so as they cling to a cosmology they themselves no longer accept all the way down. These are the enablers.

In this study I explore how one cosmology, replete with important variants, informed the initial conquest of "paganism" in what became Europe, how it helped to set the stage for a second conquest in the Americas, how it promoted anti-Semitism and the persecution of other non-Christian peoples, how it joined forces with capitalist expansion to accelerate growth and conceal climate wreckage, and how it eventually helped to reorient many white working-class dispositions from modest egalitarianism toward new fascist movements that demean and attack racialized minorities, immigrants, degrowth experiments, and climate activism. A cosmological dimension was operative in each of these processes.

The lived cosmologies of the "west" are themselves multiple. What is more, no lived cosmology is ever the sole cause of the results it encourages. Attribution to a single cause of any type rarely works. But nested inside the institutions of western capitalism, Christianity, and secularism are existential commitments to old cosmologies that help, among other things, to bolster the tenacity with which climate denialism is propagated and the persistence with which climate casualism is maintained, even during a time when experiences of climate wreckage are overwhelming. These orientations to climate in turn play into the intensities with which climate denialists and casualists dismiss those who support climate movements. By "climate denialism," I mean a strange sense that things are out of joint that is then transfigured into belligerent refusals to accept climate wreckage as human induced. By "climate casualism," I mean overt recognition that the climate is changing fast delinked from urgent efforts to fold that recognition into everyday action, research agendas in the humanities, or dramatic political movements to contend with the phenomenon. Today, climate casualism is killing us.

A lived cosmology, as already suggested, consists of operative assumptions and insistences about how the earth came into being; it also expresses a tacit sense of how time flows, bumps, or otherwise proceeds into the future. The Christian cosmology of Augustine, for instance, posits a timeless god who created the earth, followed by divinely ordained earthly processes gradually advancing toward a second coming.[3] An original sin broke the smoothness of time after its occurrence, and planetary interruptions before and after the birth of Jesus are defined to be divine punishments imposed on a sinful populace. Many western contemporaries who think they have outgrown this cosmology retain vague cultural prompts and premonitions from it that erupt when a new disturbance arises. By contrast, the pagan cosmology of Hesiod focuses on the violence by which the earth arose out of chaos followed by imperfect cycles of repetition replete with volatilities. The Aztecs, in turn, charted five phases of time in a precarious world, each ending in an earthly cataclysm, though human blood sacrifices might help to delay the next cataclysm.

To study such cosmologies comparatively is also to explore how each folds into specific institutional patterns of expectation and discipline, existential hopes, familiar stories, and spiritual anxieties. Each culturally implanted cosmology helps to shape how people respond to unexpected events that appear to call it into question. Sometimes the choice becomes stark: do you revise the cosmology with which you have been imbued or blame vulnerable constituencies for inducing or misreading the untoward event?

This study seeks to excavate western cosmologies that have tended both to promote and to conceal the climate wreckage of today, as well as other cosmologies that might help to expose the workings of that wreckage. The project is complicated by the fact that several of these latter traditions, from which enduring insights might be drawn, were muted by two historic conquests of pagans by Christian/European societies. The first conquest was the work of Christendom within what soon became Europe. The second involved massive European settler invasions of the Americas, infused with a sense of Christian, racial, and temporal superiority over peoples who occupied those lands. These conquests incurred incredible suffering; they also made it less likely that the invaders would learn much from the interlinked land practices, cosmological investments, and future expectations of the diverse peoples themselves.

The effort here is to sharpen comparative awareness of lived western cosmologies today, and then to rework them experimentally, partly in the light of old pagan insights and partly in the light of new experiences of climate wreckage. It is an exploratory adventure.

There are doubtless multiple ways to pursue such a task. A few features mark the attempt in this study. First, it focuses on a history of thinkers who advanced cosmological traditions that became prominent in the west. That approach allows one to dig into interior relations within a theo-philosophical cosmology; above all, it exposes the existential lures and anxieties that have bound people to it. For a lived cosmology is never established by detached argument alone: the existential hopes and anxieties it promotes pull people to it, even to work urgently to salvage it when events call it into question. A historical approach that encourages us to listen again to those lures can track the resilience of an old cosmology and to grasp strategies to retain it in the face of new shocks.

In the first chapter, for instance, we review the pagan cosmologies of Hesiod, Prodicus, and Ovid before Augustine enters the picture with his powerful promulgation of one Christian cosmology with its distinctive lures. Then, in chapter 2, we strive to look again at how the bishop of Hippo fashioned detailed regimens both to enforce that cosmology among the faithful and to impose disciplines over Roman pagans of the day who resisted it. Similarly, the following coda (a brief intervention between chapters designed to unsettle textual continuities across time and place) explores the feminist theologian Catherine Keller's textual identification of a Hebrew god working upon a "bubbling deep" already there. Keller locates a latent creedal affinity across differences between Judaism, Christianity, Islam, and European paganism, and she allows us to listen again to suborned themes floating in the Christian tradition itself.

Chapter 3 reviews the tortuous account by Tzvetan Todorov of the sixteenth-century European conquest of Aztecs. It then makes an exploratory foray into the complexities of Aztec cosmology itself. This was a venture Todorov himself was not really ready to pursue. The exploration may provide additional insight into what the agents of conquest and conversion were so anxious to save; it may also suggest how to draw elements from Aztec cosmology to come to terms with the shakiness of today. For the Aztec creed of a volatile, precarious world speaks to experiences today that are short-changed within Augustinian and secular projections. Similarly, the coda on Tocqueville, presented after the Todorov chapter, reviews how critical the protection of a Christian cosmology was to Tocqueville's wariness of freeing enslaved African peoples within the American continent itself and to his endorsement of settler conquests designed to erase indigenous peoples from the entire continent. I confront Tocqueville with rejoinders from William Apess—perhaps the first native North American to write in English—and Roxanne Dunbar-Ortiz, a writer who blows up the immigrant/nativist debate that continues to shape much political discourse in the United States. That coda, too, explores how an old frontier myth of implacable settler advance in North American has now morphed into fascist strategies—well symbolized by Big Lies, a wall on the US-Mexico border, and calling adversaries vermin—to close people of color out of now settled places. A movement from settler expansion to settler territorial protection, roughly within the same cosmology.

Chapter 4, on how and why Descartes and Kant defended so vehemently their respective images of subjectivity, objectivity, and the advance of singular time, is countered with an Amazonian world of diverse, intersecting subjectivities extending well beyond humanity. That multiperspectival reading is informed by the anthropologist Eduardo Viveiros de Castro. The non-Kantian image of multiperspectivism that emerges is then compared to the new perspectivism advanced by the quantum theorist Carlo Rovelli, for whom time sometimes "jumps like a kangaroo." Western, Christian, secular cosmologies of a singular line of time are shaken by such explorations.

The coda on "Nietzsche and the History of an Error" condenses themes advanced to this point, as it draws attention more sharply to existential lures that have attracted people to western cosmologies at different stages. It also calls into question the primacy of a binary logic in several of those traditions, opening the way to come to terms with more bumpy and creative moments of becoming in temporal developments. Nietzsche defines his favored orientation to be both defensible and open to contestation, calling for a "spiritualization of enmity" with respect to contested cosmologies.

Finally, in the last chapter a Bengali writer (Amitav Ghosh), a contemporary Irish geologist (Bill McGuire), and a French neo-pagan philosopher (Michel Serres) are placed into conversation to help forge a counter-notion of time as a bumpy multiplicity composed of intersecting temporalities moving on divergent trajectories and at different speeds. Diverse temporalities that once converged to support various species niches now diverge in ways that threaten innumerable species. That chapter invites readers to reject the "catastrophism" that informs right-wing ideology and covers up real potential catastrophes in favor of a measured catastropheism that comes to terms with the worst dangers of today. It closes with discussions of how today "impersonal planetary circuits" distribute the worst initial effects of climate wreckage launched in temperate zone, capitalist states to polar, semitropical, and tropical regions which release the lowest levels of carbon into the atmosphere. And it revisits the improbable necessity of a politics of swarming to cope with today.

The image of time as multiplicity is better suited to come to terms with the modern time of climate wreckage. But it is less equipped to sustain the lures that have drawn Euro-American settler societies for centuries to old Christian/secular cosmologies. The promise of an afterlife, or indefinite material progress, or continued profit opportunities, or all three, are placed into jeopardy by such a cosmology. Are there new or revised lures to draw upon today? That critical issue, too, is joined, if hesitantly. For the old existential lures no longer resound with a ring of confidence. And right-wing belligerent reactions to those very disappointments in old promises of the future become more dangerous by the day.

But how does a mere political theorist come to terms with such historically diverse strains of thought and practice, several outside his previous purview? Moreover, should anyone really attempt such a sweeping intellectual strategy? One response to the second question resides in the complex relation between lived cosmologies of the west and the latest onslaught of climate wreckage. Many cultural theorists need to pursue these heterogeneous and volatile interconnections. Perhaps a reply to the first question is that political theorists, because of their very precarious position between the humanities and social sciences, are often among those willing to address the diverse fields required to cope with such a multifarious event. At any rate, in cases where I wander into traditions heretofore unfamiliar to me I seek sustenance from scholars who specialize in these domains. To take one instance, scholars such as Camilla Townsend, Frances Berdan, and James Maffie inform my reading of Aztec cosmology. I then pursue comparisons with earlier European pagans in order to stretch this adventure beyond the range the scholars themselves may pursue. How might images of temporal processes less highly predisposed to human beings inform new images of time in the west today?

A critical intellectual, to me, is one who pursues the logic of a problem where it takes them. The central problem I pursue is the time of climate wreckage and lived western cosmologies that helped, first, to promote these conquests, second, to sow the seeds of climate wreckage, and, third, to conceal it from widespread attention in settler societies until late in the day.

To pursue a problem where it takes them requires explorers to work diligently within the zones in which they are most competent and then to range into others, as the dictates of the problem itself require. In this text, that involves both folding work by several recent geologists and earth scientists into the humanities and social theory themselves and exploring nonwestern cosmologies that help us better to discern problematic existential demands that still sustain western cosmologies. Both. Amateurs wander, under the star of the problem that perplexes them. They must also, of course, welcome corrections and improvements when their wandering takes them into unfamiliar territory. But a critical intellectual must not allow such difficulties to deter them from undertaking such forays themselves, when needed. Such is the role of the critical intellectual during a time of climate wreckage. Neoliberal universities encourage a blander interdisciplinarity than is needed today. This is a time when transdisciplinary adventures pull the humanities and planetary sciences closer together.

Why did it take so long, within temperate zone, capitalist states, to come to terms with the climate wreckages they initiated? Sure, greed, glory, and the pursuit of new profits played roles of great significance. But those activities themselves also gained prominence and legitimacy in part from the cosmologies in which they were set. Can comparative explorations of lived cosmologies further help to show how this was so? Can they even help set the stage for renewed cosmological explorations and better strategies to respond to climate wreckage? My answer to such questions is a shaky yes, for I have become more sensitive to how fervent desires to protect old cosmologies feed into the intensities of climate denialism and the self-deceptions of climate casualism.

To gain a preliminary sense of how old cosmologies delayed recognition of climate wreckage until it was well under way, let's take a look at the lingering terms of environmental discourse in the United States during the 1970s. Two competing positions dominated the landscape. Capitalist and productive socialists, while differing fundamentally on how to do so, supported political economies of rapid growth in material abundance. Progress was the name of the game, and it was tethered to economic growth. These drives faced increasing resistance, after a huge oil spill in 1970, from growing movements to become more ecological and roll back growth imperatives. You could identify Keynesians, most Marxists, and Hayekians in the first capacious camp, and theorists such as Barry Commoner, James Baldwin, Mary Daly, Charles Taylor, Jürgen Habermas, and Fred Hirsch as exemplars of the second.[4] The first side, particularly the capitalists, held by far the most institutional clout; the second made impressive intellectual inroads and supported various critical movements.

I became attached to the second movement early in my intellectual career. But both contending movements remained for too long unalert to rapid climate change occurring in the oceans around them, the atmosphere above them, distant glacier flows, and the ground below, even though a few dissenting voices in neither of these camps did try to highlight those processes.

So, at least two opposed western images of modern life on earth: economic expansionists and egalitarian earth lovers. But by the early 1980s, two unexpected events disrupted *both* images of actuality and possibility together. First, the growth-mastery agenda, often inflected by Keynesianism, was quickly overmatched by a white evangelical/neoliberal constellation that pushed things toward the far right. Rapid growth was now to be promoted through a neoliberalism that distributed wealth and income even more toward the upper one percent; an evangelical image of divine, progressive time, linked to repression of diverse minorities, soon became woven into this movement. Critiques of the state developed by the new earth lovers were hijacked by an angry coalition organized around white racism, anti-welfarism, militia organizations, Christian imperialism, high-roller donor groups, renewed demands for gender hierarchy, and subterranean strategies to lock in Republican control over the state so that no critical movement could succeed.[5] The grievances of the white working class were significantly, though not entirely, absorbed into the new neoliberal/evangelical constellation. It was not a constellation devotees of eco-legitimacy had anticipated. One key to the new constellation is how its two major constituencies intensified the demand to sustain a cosmology of singular time, drawing capitalist progress and the promise of eternal life closer together again. As Weber says they were during an early advance of industrial capital in northern Europe.

A second development went unnoted in established research circles at first. Capitalism itself had become closely imbricated with a host of large, volatile planetary processes that escalated climate warning. Complementary assumptions governing debates between the pro-growth/mastery and the pro-earth/harmony parties helped to screen this unfolding phenomenon from wide scrutiny and exploration. A shared cosmology distracted each from attending seriously to old folk stories about how planetary processes periodically become volatile. The assumption of *planetary gradualism*, at that moment shared by most western humanists and geologists, contained remnants of an old western cosmology of earth time as rather smooth and slow. Those attached to various versions of such a cosmology were ill prepared to perceive distinctive things happening on the ground, in the atmosphere, to the ice sheets, on the deserts, and within the oceans. The diverse instruments, technologies, sciences, phenomenologies, existential faiths, and observational powers they deployed did not point them toward such developments.

Most English and American geologists and climate scientists embraced the assumption of planetary gradualism that Charles Darwin, the nineteenth-century evolutionary theorist, and Charles Lyell, the renowned nineteenth-century geologist, had perhaps unconsciously absorbed from Christian traditions and set into the stone of theory. That is the assumption that planetary temporalities—including species evolution, the organization of oceans, the evolution of the atmosphere, the pace of climate change, the tempo of glacier flows, the reliability of monsoons, the size of deserts, and so on—change on long, slow time. According to this premise, it is safe to assume that the basic contours of the geological past roughly correspond to the shape of the present. And yet alternative cosmologies, muted earlier by two conquests of pagans, could have alerted more humanists, geologists, and social theorists to other possibilities. I hope it is now apparent, though it will become clearer soon, that I am not saying new intuitions were impossible at the time in question. Rather, I am saying that they were less likely to be formed under the unconscious influence of such cosmologies and, when so, even less apt to be taken seriously by professional constituencies.

These were not just assumptions and existential lures, then, shared by most western geologists, oceanographers, climatologists, and evolutionary theorists late into the 1970s; they were also absorbed differentially into the collective unconscious of renowned Euro-American humanists, theorists, phenomenologists, theologians, economists, secularists, and philosophers. Sigmund Freud, Max Weber, Louis Althusser, John Rawls, Charles Taylor, Jürgen Habermas, Hans Reichenbach, Fred Hirsch, Bertrand Russell, Hannah Arendt, Friedrich Hayek, John Maynard Keynes, Martin Heidegger, and Paul Tillich—each diverging from others in significant ways—roughly shared complementary planetary, cosmological assumptions. Some tended toward a cosmology suited to human mastery over the earth, while others were drawn to one in which human beings could enter into close attunement to a harmonious earth.

Freud tended to reduce Sophoclean dramas to a family Oedipal triangle, paying little attention to real, virulent plagues and rocky planetary events coursing through the dramas. Such earth disruptions were widely treated by literary gradualists more as symbols of cultural struggles than as natural realities. Althusser, at this stage, insisted that the economy rules "in the last instance," seemingly unalert to how massive, rapid climate changes had aided, threatened, or destroyed entire civilizations in the past and were under way again. Heidegger worried immensely about how technology permeates modern modes of temporal experience and carries dark possibilities, but he did not fold larger planetary volatilities into those conjunctions. Charles Taylor, at this juncture, explored how to harmonize more closely with the earth, an earth not composed of multiple temporalities that periodically fall out of sync with one another. Habermas did voice a dark suspicion about possible "climate heating" in the German publication of *Legitimation Crisis* in 1970, but he did not discern how a checkered history of planetary induced climate changes could help to explain how this one, too, was unfolding. His gradualism was punctuated only by the interruption of the "Anthropocene," a term that itself did not come into use until 2002.

3.

It is not as though counter-voices were lacking on this score, either at the time or before. But they suffered under a disadvantage. Cuvier, the great French naturalist of the late eighteenth and early nineteenth centuries, had already identified a series of species "catastrophes" suggested by huge gaps in the fossil record. But his claims were buried under the charge of being "mythic" by Darwin and Lyell. The folk wisdom of many pre-Christian European and American pagans, as we shall explore in greater detail in this text, often appreciated tumultuous planetary processes of numerous sorts, frequently personified by gods. Such images were discounted, marginalized, or silenced by the combined authority of Christian and secular voices in the west, voices that overlooked affinities they shared with their favorite debating partners.

Consider merely a sampling of folk reports about climate changes from earlier times. "According to traditional Chinese chronologies, the rise of the Xia dynasty is often dated to 2070 BCE and linked to the devastating flood."[6] A pharaoh in 2180 BCE reported severe famine linked to a long drought along the Nile. The biblical report of the flood of Noah, long treated as a mere fiction by secularists (such as me), has recently been accepted as a reality by many geologists. Ovid, as we shall see, also recorded it. Flood legends in India, among

American indigenous peoples, and elsewhere were also common. Reports circulated in Rome of the first century CE about a warming period that allowed grapevines to grow higher in the hills. Reports in Rome circulated in the late 500s of rapid cooling, associated with a plague outbreak. Amerindian knowledges adjusted to what is now called the Little Ice Age on the east coast of America (ca. 1300–1850 CE). Chronicles charted the expansion of farming zones in Europe of 1000 CE during what is now called the Medieval Warm Period. In the late Middle Ages, reports arose of "winters of such unaccustomed severity and depth of snow that many people died." Again, in what we now call the Little Ice Age, reports appeared of mountain glacier expansions that blocked old roadways in northern Europe. And there were reports in the 1690s in Scotland about "six consecutive seasons of disastrous weather when the harvest would not ripen."[7] These chronicles, stories, and reports, especially when emanating from pagan authors or peasant chroniclers, were often held later to be mere fabulations.

Additionally, by the 1960s, marginal western climate scientists began to talk about significant shifts in climate over the millennia. There was even a *Time* magazine article in 1956 quoting an oceanographer who said that in fifty years carbon emissions "would have a violent effect on the earth's climate."[8] But that report was buried under the Cold War, coal and oil interests, popular desires for capitalist progress, a focus in physics on the intricacies of relativity, and the vague lure of unilinear time on the way to progress.

One such climate science dissident was Hubert Lamb, who identified earlier periods of severe, sometimes rapid climate changes. If you scour his books, you find him wondering how dinosaurs disappeared so fast, though he could not find a plausible reason. He identified what are now called the Medieval Warming period and the Little Ice Age. He even marked the Eemian period, now said to have occurred 130,000 years ago, an extremely hot planetary era of (what he thought was) 11,000 years in duration. Though, again, he could not locate its causes. By 1970, he concluded that climate warming would escalate in the near future because of the extensive CO_2 emissions emanating from fossil fuel use in cars, construction, electric generation, and cement production. But he was a rather lonely voice.

This remarkable scientist, I sense, was poised between what we now call geo-gradualism and later appreciation of periods of rapid change in climate, oceans, glaciers, deserts, evolution, and species habitat. He discovered periodic changes that were rapid and deep. But he also strove to keep those findings consistent with "uniformitarianism" (what I call gradualism). "In making such a diagnosis, the meteorologist will be seen to be adopting the geological principle of uniformitarianism—that 'the present is the key to the past.'"[9]

His observations began to rattle established views, but the planetary paradigm to which even he confessed allegiance made it more difficult to pursue those findings robustly.

What if, while exploring old phenomena with the rough instruments then available, Lamb had also consulted a series of pagan cosmologies that assume the planet goes through periods of volatility, with multiple stories, myths, tragedies, and philosophies of wisdom set within such understandings? Could a to-and-fro movement between his work and those cosmologies have freed this exploratory scientist more? I do not, of course, know. The barriers to such engagements would have been severe, however, since pagans carried little legitimacy as sources of possible insight and wisdom in prevailing circles. As things stood at the time, Lamb's findings did not make much of a splash in his own field, and none at all in the human sciences, the humanities, and the public at large.

The next part of this story is now widely known. In 1980 Luis and Walter Alvarez presented indirect evidence that a huge asteroid had hit the Yucatan Peninsula 66 million years ago, quickly wiping out all dinosaurs and a huge percentage of other land and sea animals. The old story of planetary gradualism within geology and paleontology now began to move onto the defensive, though an intense debate on the issues raged for at least another decade.

By the 1990s several mass species extinctions had been uncovered, each with distinctive planetary sources. A new geological science based on long periods of slow change punctuated by shorter periods of rapid turns now came into its own.[10] Those shifts were roughly correlated with another series of changes in the earth sciences that emphasize the proliferation of nonhuman agencies in the world, with viral species crossings underlining how such diverse agencies can surge into human cultures. The problem now became how to bring a history of such a periodically volatile planet and more than human agencies to bear on new evidence of climate change today, and how to do so in ways that both create a more dynamic theory of climate wreckage and expose how radical and rapid counteractivities must be to counter its worst effects.

4.

Another maverick who preceded the Alvarezes, the oceanographer Wally Broecker, can now be wheeled onto the scene. His work not only transformed earth science understandings of oceans, it also helped set the stage to rethink lived western cosmologies. Broecker first discovered the global ocean conveyor in the 1970s and soon showed how it had shut down rapidly 12,700 years ago, though many earth scientists discounted those latter findings for a while. The worldwide ocean conveyor 12,700 years ago was stymied at its most vulnerable junction. The downward spiral of cold Gulf Stream water around Greenland was blocked by the rush of salt-free (and therefore lighter) water from melting glaciers at the junction where salty water had previously driven down, to flow south on the sea bottom. Broecker soon concluded that a similar process is under way again, even though many of those impressed with then extant climate computer models disagreed with his findings.[11] Now, most oceanographers agree that such a weakening is again well under way and a new closure is very apt to occur.

How does it work? And how suggestive are these workings for cosmological thought? Well, accelerating capitalist CO_2 emissions first heat up the atmosphere; the warming Greenland glacier then enhances ice melts; the conversion of ice into water increases the absorptive capacity of the sun (the albedo effect), instigating a self-amplifying loop of new melts; formation of huge crevices in the melting ice then allows water to rush to the bottom to grease further the pace of the flow; uplift induced by reduced glacier weight then spawns new earthquakes that further amplify the pace of glacier flow and resultant iceberg calvings. The accelerating calvings flood the sea with fresh cold water. At least five amplifiers bouncing and seeping into one another, then, in the aftermath of the carbon trigger.

Now a counterintuitive effect is produced. The mass of cold, salt-free water released into the ocean by climate heating is lighter than salty water heading north that had heretofore been driven to the bottom as it cooled. So the Gulf Stream momentum is slowed, and could well stop in the future, as it did in the recent geo-past. The dynamic result could be a closing of the ocean conveyor again. The closing will generate another cold regime in the eastern United States and western Europe, while the now stagnant conveyor spawns even more heating than otherwise in tropical and semitropical zones.

That is one example, with others to come, of how climate cascades occur, with one driver triggering an amplifier, and it yet another, until the composite result is both much larger than the initial trigger and sometimes at odds with prior intuitions. The forces in such a cascade not only impinge upon each other, but many also ingress into others. Moreover, they often set off impersonal planetary *distributors* which carry the worst initial effects from temperate zone emitting states to tropical, semitropical, and polar zones.

Such trigger/amplifier cascades, as they pour differentially into human and nonhuman cultures, require those in the humanities and social sciences to acquire a more vivid awareness of the rocky planetary history in which they participate, paying attention to diverse periods of volatility and how previous times of climate heating are apt either to vary or to be repeated today. Most of the issues humanists explore—including racism, immigration patterns, decolonial struggles, the emergence of fascist movements—are imbricated with these planetary processes. We will encounter such dynamisms, as we chart cascading linkages between capitalist high-growth economies, carbon emissions, planetary amplifiers, impersonal planetary distributors, warming trends, intensified droughts, wildfires, devastating floods and storms, monsoon interruptions, more intense and extensive human migration drives from south to north, and possible neofascist responses by old, temperate zone, capitalist states to these very cascades.[12] Cascades in which each event helps to trigger other heterogeneous events, but also in which the reserves and creativities of the affected parties mean that the triggers do not always determine the results.

Cascades of impingement and infusion into diverse entities. The last chapter of the book argues, with help from others, that close attention to how multiple temporalities periodically intersect—cascading temporalities lodged in capitalist triggers, diverse planetary amplifiers, and militant cultural responses to both by constituencies clinging to the lure of old, western cosmologies—*suggests the need to rethink time itself as a multiplicity shaped by intersecting temporalities set on different trajectories, capacities, and speeds.* Viral temporalities, bacterial temporalities, civilizational temporalities, glacier temporalities, monsoon temporalities, ocean acidification temporalities, earth heating temporalities, plant temporalities, human migration temporalities, acidification temporalities, aquifer temporalities, fungal temporalities, and fascist movement temporalities—each intersecting from time to time with others and shifting in response to such intrusions. Moreover, during a time of climate wreckage the breakup of old temporal symmetries between, say, parrots, insects, larvae, and hurricanes in Puerto Rico can decimate the parrots. There are innumerable examples of how previous temporal symmetries become asymmetrical today, often with devastating effects.[13] Such asymmetries have unfolded before, to be sure, but the speed of climate wreckage today decreases the time various species have to adjust to emergent asymmetries.

This book, then, pursues a counter-introduction to some major traditions of western thought; it is designed to chart key omissions, cruelties, and blind spots in those traditions, focusing particularly on cosmological assumptions and existential demands that helped to foment and secure them. An image of time as singular and linear, for instance, grows out of most monotheistic and secular European cosmologies.

The task is not always to blame Euro-American parties for omissions that hastened the current time of climate wreckage. That only works up to a point and, indeed, I was one of those parties. The larger agenda, rather, is to address how and why such blind spots persisted so long in dominant traditions of western thought, even though engagement with several subdued pagan traditions inside and outside Europe might have helped to short-circuit those assumptions and the existential lures attached to them. The focus, then, is on existential demands and intellectual assumptions that have sustained several lived cosmologies of the west, setting the stage in the last chapter for consideration of another competitor to Christian/secular cosmological, strategic, and existential dispositions.

Capitalist imperatives, the greed of venture capitalists, fossil fuel company land invasions and charades, Christian pursuits, and obdurate consumer practices are not to be discounted in such an exploration.[14] But existential orientations to time inhabiting these very cultural practices also need to be exposed and reassessed. Such assumptions and lures have intensified various diversions, legitimized practices of climate wreckage, and tended to obscure the results from the perpetrators. Choices have to do with relatively banal options; decisions with fateful existential turns. What appear to be choices under one set of assumptions can later be exposed to be fateful decisions.

5.

The chapters in this text follow a regular historical order. Each of the first four chapters, as already noted, is then punctuated by a *coda*. A coda, often set in a different time and place from the chapter preceding it, interrupts the historical flow in something like the way an unexpected natural event can disrupt a temporal trajectory. A coda might draw upon a thinker from another time or place to accentuate a theme already in play, or to challenge one that had provided a focal point in the theo-philosophy under review, or to offer a condensation of what has gone before. The *codae* on Jocasta and James Baldwin, Catherine Keller, Tocqueville, and Nietzsche thus serve several functions. Jocasta and Baldwin alert us to how a tragic image of possibility can jump from one time to another. The Tocqueville coda exposes the obduracy with which Tocqueville sought to protect his cosmology on an unfamiliar continent. And the condensed "history of an error" by Nietzsche highlights several lures that have attracted so many to both a binary logic and a linear cosmology. It even challenges the lures of a positivism making inroads again today among venture capitalists, social scientists, and university presidents.

It is thus sometimes productive, as you examine the context in which this or that strain of thought is set, to jump into another time or place to gain a fresh perspective on it. Doing so to reopen shuttered horizons of thought and to untether stubborn existential aspirations. For an existential inspiration sometimes leaps across time and place, as expressed in Baldwin's debt to Sophocles and that of Michel Serres—a European philosopher of science and time examined in the last chapter—to pagans such as Archimedes and Lucretius. Major figures in western thought themselves have often taken such jumps across time, as Hobbes did with Job, Augustine with Plato, Kant with Epicurus, Descartes with Augustine, Arendt with Aristotle, and Rawls with Kant. The Muses I draw upon in this text merely differ from several of those informing major western traditions.

6.

Yet, it may still be asked, why the term "pagan"? Why introduce such a charged word as a term of art into this text? According to *Merriam-Webster's Eleventh New Collegiate Dictionary*, *paganus* in pre-Christian Rome meant a rustic country dweller. With the rise of Christianity it became a yet more derogatory term of description. "1. Heathen; *esp.*: a follower of a polytheistic religion; 2. one who has little or no religion and delights in sensual pleasures and material goods." Augustine, as we shall chart, reviled Greek and Roman pagans of a nonplatonic cast. They populated the world with multiple, dirty, copulating, cruel deities. And they did not look forward to either a second coming or, very often, a beautiful life after death. His extensive condemnations of them in *The City of God against the Pagans* vindicated their replacement by Christendom and, indeed, helped to define an emergent Europe. It also helped set the stage for a later conquest of the Americas.

This book, then, constitutes another attempt to lift the term "pagan" out of the mud. It does so, first, by extracting elements from early Greek and Roman cosmologies obscured and demeaned beneath the combined weight of Euro-American Christian and secular imperialisms. Elements that—particularly when joined to belated turns in western thought itself to acknowledge the vitalities of the nonhuman—are replete with insights pertinent to a new time of climate wreckage.

Climate ravages, western imperialism, pagan appreciations of multiple vitalities in the world that include and go beyond the human, and conceptions of time embedded in capitalism and Christianity must be explored together. The point is not only to place some pagan modes of thought into critical relation with what became major Euro-American traditions. It is also to mark selective affinities between early Greek/Roman modes of paganism and later peoples conquered by European Christendom in the Americas. Affinities, not identities.

The first attack on paganism in what was later called Europe was less racial and more cosmological. The second imperial, territorial attack was both extremely racist and highly cosmological. Both conquests also involved subordination of women, gender duality, class divisions, and harsh strategies of land ownership and exploitation. Deforestation and the transformation of open landscapes into private parcels of land subjected to intensive agriculture, for instance. The intensities of those attacks on pagans involved, among other things, urgent attempts to save images of earthly time and the possibility of an eternal afterlife from lived pagan counterexamples of how to be and feel in the world.

The two conquests, certainly, do not explain everything about the current time of climate wreckage, a time many continue to experience with carefully cultivated casualness. Nor do I suggest that earlier pagan cosmologies perfectly prepare people to come to terms with today; rather, several such cosmologies *help* to account both for formation of climate wreckage and for the existential screens that blocked attention to it until late in the day in settler societies. Climate casualism, still widespread in several western states, is one effect of that complex. Indeed, a little climate casualist continues to subsist in most of us.

To say that pagans have been territorially conquered twice is not to say, either, that such voices and cultures have been eliminated. That would both be false and give those outside those traditions little sustenance to draw upon. It is to say that such voices have been muted and delegitimized in major Euro-American traditions. Pekka Hämäläninen, for instance, recently completed an impressive study showing how long and creatively multifarious indigenous peoples have fought to retain their lands and cultures in North America, with impressive success. Ned Blackhawk has completed a yet deeper study to accomplish that task.[15] Amitav Ghosh, addressed in the final chapter, explores how indigenous modes of vitalism have been "muted" but not eliminated through European conquests. This text profits from those inquiries as it seeks to pull some devalued themes more securely into contemporary European thought.

To challenge dominant western cosmologies it is also necessary to disturb and work upon a larger complex of orientations entangled with them: the subject/object dichotomy, the primacy of binary logic, human exceptionalism, planetary gradualism, private parcels of agricultural land, sociocentrism, capitalist imperatives of growth, and eminent over emergent modes of causality, among them. The task is not to show how the above themes are locked together in tight *chains* of causality. The counter-cosmology pursued here eschews the simplicity of such chains. Rather, such formats and practices are loosely intertwined, with one suggesting another and it yet another, until a *tangle* of practices, assumptions, and lures becomes consolidated. It is perhaps by addressing a series of pivotal moments in western thought and imperial conquests that such complex patterns of entwinement are best illuminated. At least, that is a wager of this text.

One other question. Is this, though, really a series of critical engagements with *western* thought? Well, I concur with Michel-Rolf Trouillot: "The West does not exist. I know, I've been there."[16] It does not *exist* as a unity because its historical fabrication required the defeat and silencing of vibrant modes of thought and practice inside and outside its imposed territorial complex. That fabrication included political formation of whiteness as a key political identity. One aspiration of this text is thus to examine a set of "western" classics in relation to lived orientations that had to be silenced, delegitimized, marginalized, or demeaned for them to achieve hegemony over the territory in question: To become the west. To enact such a critical strategy is to shake the myth of the west. I know, I've been inhabited by it.

The final chapter of this study draws these threads and themes more obdurately into the contemporary world. It places Amitav Ghosh's recent account of the devastating effects of European imperialism on colonial peoples, their terrains, and the planetary atmosphere, first, into conversation with geologist Bill McGuire's account of how the time of severe climate wreckage is now "baked in"; and, second, with philosopher Michel Serres's judgment that it is now timely to acknowledge time itself to be a bumpy multiplicity composed of intersecting temporalities moving at different speeds, trajectories, and capacities. Ghosh perceives how various modes of "vitalism" in pagan and other non-Christian societies have been demeaned through conquest but not destroyed; the earlier chapters in this text are extended by Ghosh to Euro-imperial drives in India and the isles of the Indian Ocean. Serres, identifying himself as a neo-pagan philosopher, carries the Ghosh account of multiple vitalities and forces in the world into a reconstitution of western images of time itself. To place a western geologist, nonwestern anthropologist, and post-pagan philosopher into conversation then allows us to explore how "impersonal, planetary circuits of imperial power" become organized today, adding climate insults to imperial and class injuries already addressed in decolonial theory. It also encourages us to rethink time as a multiplicity, a task I carry a step or two further in the light of a long history of planetary volatilities.

It is doubtless difficult, these days, for those who care about the multifarious effects of rapid climate destruction, to ward off depression and a sense of hopelessness. I certainly have to work to do so. This study closes by revisiting ways to appreciate the grandeur of an earth with a life of its own and to activate "the improbable necessity" of militant climate activism.

1
Hesiod, Ovid, and a Turbulent Cosmos

Hesiod and the Cosmos

Hesiod, a peasant farmer in ancient, patriarchal Greece, collected stories about the gods and presented them in prose poetry to the Greek world in the *Theogony*. It was because the Greeks had recently borrowed the Phoenician alphabet of consonants around 850 BCE, and an anonymous Greek had then added five vowels to it, that such a book could be written in a script available to a broader readership. Literacy was not yet extremely common, but the new script created a more expansive audience.[1]

The book itself was composed around the eighth century before the current era (BCE). It bears the imprint of Hesiod, who, unlike the anonymous author of the *Iliad*, both named himself as author and was less enamored with heroic virtues. He was more concerned with how peasants could make a go of it in a tempestuous world. Life could easily become rattled and unsettled: by exuberant or angry gods who personified and activated various forces of nature, by conflict between polities, by rebellions within a regime, or by a confluence of such diverse forces. Life was thus filled with uncertainties and contingencies, some emanating from the turbulent character of the nonhuman world, others exacerbated by it. And joyous blessings could be distributed by the same means. The gods were often feared and propitiated, even more than worshipped as benevolent, caring beings who placed humans at the top of their concern.

Hesiod's text begins when he asks the Muses to guide him in collecting and telling stories about the gods. "They breathed into me their divine voice, so that I might tell of things to come and things past, and order me to sing of the race of the blessed gods, who live forever. . . . But enough of this gossiping."[2]

The gods occupy and energize a cosmos that exceeds them. Void, or Chaos, precedes divine creativity. "Out of Void came darkness and black Night and out of Night came Light and Day, her children. . . . Earth [Gaia in Greek] first produced starry Sky, equal in size to herself. . . . Next, she produced the tall mountains, the pleasant haunts of the gods, and also gave birth to the barren waters, sea with its raging surges. . . . Thereafter she lay with Sky and gave birth to Ocean with its deep current."[3]

Already we encounter a cosmos irreducible to later Christian images, as well as to yet later secular readings that both challenge and grow out of Christian readings of the cosmos and its manifold relations to humanity. Christian stories typically set the origin of the universe itself in the creativity of an omnipotent god. God even creates time and places humans far above other beings, for the latter lack souls. Secular revisions of that story often set the evolution of life on earth in long, slow processes that culminate in humanity. Hesiod deviates from both of those stories.

It all, to Hesiod, *begins* in chaos.[4] The emergent, cosmic order soon contains conflicts of desire, or purposive striving between the gods who inhabit tall mountains, barren deserts, raging seas, and powerful ocean currents. While it is not perfectly clear when humans emerge in this story, once they do the cosmos provides the multifarious world in which they strive and with which they contend. Notice, too, the political language through which the cosmos is portrayed. Quarrels, strife, imbalances of power, wars, raging desires, treachery, and contingencies inhabit it. It is also a world saturated with fabulous sexual desires, jealousies, and contests. Consider a few instances.

Cronus is born of a cruel father who sees his son as a dangerous competitor for hegemony. So, the mother of Cronus, Earth/Gaia, urges Cronus to kill his father. "Mother, I am willing to undertake and carry through your plan. I have no respect for our infamous father, since he was the one who started using violence."[5]

Enormous Earth/Gaia hid her son and gave him a huge sickle. "Then, from his ambush, the son reached out with his left hand and with his right took the huge sickle with its long, jagged teeth and quickly sheared the organs of his own father and threw them away, backward over his shoulder."[6] The first divine ruler, never all that secure in his rule, is castrated and overthrown by his son. It might be enough to make Freud quake in his boots.

But Cronus and his retinue of Titans—all huge gods—are also violent, and Earth/Gaia promises that they too will face a similar fate someday. Eventually Zeus—the son of Cronus who also needed a ruse organized by his mother Gaia upon his birth to escape the wrath, or homicide, of his father—begins to assemble a force to fight Cronus. The troops are recruited from young, discontented

Titans, the set of gods ruled by Cronus. It seems that mankind had come on the scene at least by the time the war between the Olympians and the Titans begins. Along the way, Zeus punishes the god Prometheus for taking pity upon humans and giving them fire to help them through harsh winters. And then the war between the Titans and Olympians erupts; the Greeks can hear and feel the battles in the lightning and thunder of the sky, in the swelling of earthquakes, and in the raging seas:

"The mighty Titans fought from the top of Mount Othrys, while the Olympian gods . . . fought from Mount Olympus. For ten full years they fought without ceasing, so bitterly did they hate each other."[7] When Zeus adds new allies, the Titans do the same. And the celestial war rages with new intensity. Lowly humans can hear and feel its terrifying echoes, rumbles, tremblings, and groans. "The limitless expanse of the sea echoed terribly; the earth rumbled loudly, and the broad reach of the sky shook and groaned. Mount Olympus trembled from base to summit . . . and a heavy quaking penetrated to the gloomy depths of the Tartarus."[8]

"Echoed terribly," "rumbled loudly," "shook and groaned," all these shattering sounds and jolting eruptions penetrate into trembling humans, in the way that being in the path of a massive tornado shakes and rattles your whole being. These traumas become collective, communicated through visceral expressions and stories to an entire culture. Even to hear over and over the story of these divine battles is to become transfixed and rattled. And perhaps amazed. Did the Greeks make all that divine noise and celestial turbulence up? Or did they deify a set of planetary volatilities that periodically interrupt the Greek world itself? That debate soon intrudes into Greek culture itself. We now know that Greece sits near the conjunction of three tectonic plates, differing in weight and consistency: one plate heading north from Africa, another east from Asia, and a third south from the Aegean. So it was (and is) the site of periodic earthquakes, volcanoes, raging seas, and tsunamis. We also know that numerous earthquakes, volcanic eruptions, tsunamis, and plagues disrupted civic life later in Greek cities. The biggest volcano—one of the most destructive in the history of the world—exploded on the island of Santorini (then known as Thera) in 1500 BCE, destroying the most robust culture of the day in the Greek archipelago. People would record that event in lore, and some could see its physical remains centuries later on the island itself. Plato even said the event occurred because the gods punished the island populace for being too hubristic. Planetary volatility accompanies the instantiation of Greek gods.

At any rate, Zeus and the Olympians eventually triumph in the cosmic battle with the Titans, and, on Hesiod's reading, their new governance of the heavens brings more stability and justice to them. But the world continues to be inhabited by elements of wildness and the periodic impetuousness of a divine ruler who neither is totally in charge nor places human beings at the top of his list of concerns. Zeus marries his sister, Hera, who becomes insanely jealous of his extramarital sexual exploits. He soon embraces Semele, a human, and she gives birth to Dionysus, the god of drunkenness, a trickster, supporter of seasonal revolts by women against patriarchy, and harbinger of bouts of wildness in the cosmos. Semele herself is soon lifted to divinity, the sort of metamorphosis from human to god that Augustine later found so despicable about the Greeks. Of course, he himself had to struggle with the issue—and internal Christian debate—of whether Jesus was born into the world divine or later became divine when his earthly ministry was consummated through crucifixion.

Other gender battles are discernible in the heavens too. Metis, the goddess of wisdom, strives to bear a daughter, Athena. But Zeus hijacks this pregnancy and bears the daughter from his forehead. The hijacking, as Mary Daly shows, continues the misogyny of the gods, as the goddess of war becomes a "daddy's girl." The god of power continues to exercise hegemony over the goddess of wisdom.[9]

In *Works and Days* Hesiod pays much more attention to the lives of mortals, particularly to the conditions of peasant farmers, and most particularly to the life of his younger brother Perses, who is lazy and needs guidance.[10] The book is addressed to him. If from the perspective of the heavens the order of the cosmos has become more stable upon the advent of Zeus, from the perspective of peasants the weight of the nonhuman world carries considerable volatility and uncertainty. Both the perspective from above and that from below are needed, hence the two different books, *Theogony* and *Works and Days*.

In *Works and Days* Hesiod summarizes five stages mortals go through, with the tendency of each stage to face more troubles and uncertainties than experienced by the "race" preceding it. There is the golden age, the silver age, the bronze age, the age of heroes, and the iron age—the last being the age Hesiod himself is said to inhabit. The age of heroes is, perhaps, an interim age in which the heroes are intermediate between men and gods. Gradually, the gods draw further from mortals themselves, and that is definitely the case by the time the iron age is reached.[11] That withdrawal is pertinent to bear in mind when you consider the age of iron.

In the early discussion of the age of iron, Hesiod suggests to Perses that those who practice piety to Zeus are apt to be most fortunate in life. But later, the relationship between piety and good fortune appears to be more uncertain. Despite the piety of peasant farmers, there can be rain at the wrong moment and drought at the wrong time, too, putting crops at risk. A plague is always possible. Or the seas might become turbulent, even raging, when your crops need to be sailed to market. Thus: "Men can sail with safety, for then a ship will not be shattered, and the sea will not wipe out the crew, unless this is the will of Poseidon who shakes the earth, or Zeus, king of the gods, wants you destroyed: both have power over good fortune as well as misfortune."[12]

It seems that "Zeus has a design for each occasion, and mortals find this hard to comprehend."[13] Under such circumstances, is pious trust in the gods the best course? Or is it wiser to remain wary of these divine activists in nature in practice while placating them in public? Hesiod prefers the first course, but the *Theogony* does not resolve the issue.

As Marija Gimbutas shows, the stories of Minoan gods that preceded the *Theogony* may depict a more matriarchal cosmos than the patriarchal renderings of Hesiod and Homer. Goddesses such as Hecate, Demeter, Artemis, and even Athena before the militarist rewriting of her in the *Theogony* are full of fecundity and care. The masculinization of the cosmos, with its repeated scenes of rape, probably symbolize a violent shift to patriarchy in these cultures.[14]

Is it possible to oppose a patriarchal image of the cosmos while respecting the periodic moments of earthly turbulence and volatility in these stories? Such a crucial task seems to call for both continued intellectual vigilance and (what might be called) aesthetic gymnastics. We will explore the latter in later sections of this text.

Greek Attempts to Naturalize the Gods

Was it possible to naturalize the Greek gods, that is, to attend to some of the volatilities and transfigurations they exhibit without investing divinity into the active features of the nonhuman world? Is it, too, possible for us today to resist the patriarchal drives inserted by Hesiod into the cosmos itself, as we come to terms through him with a nonhuman world that passes through multifarious bouts of turbulence? Is it feasible, indeed, to endow respect upon Hesiod because he personifies these gods without celebrating the heroic, military virtues that prevail in other Greek texts such as the *Iliad*?

Questioning the veracity of belief in gods was risky in ancient Greece; it was apt to encounter sharp opposition from peasants who sought to court favor with the gods and from aristocratic rulers, too, who found popular piety provided an important support for their rule. Nonetheless, merely a few centuries after Hesiod wrote, some "Sophists" began to pursue that very trail. Democritus did, for instance, though his voluminous texts are now available only in fragments. It seems the gods did not favor the preservation of texts that did not favor them . . .

Anaximander, living in Melitus (in modern Turkey) between 610 and 546 BCE, called the gods into question too. His city of origin sat at the juncture between Greek, Egyptian, Mesopotamian and Persian influences, so his thinking is not easily reducible to a purified notion of "Greek" thought. He apparently held not only that the earth is a sphere that floats in space but also that there is no law that says everything falls down, the view that had made the first claim about an earth poised in space less tenable to many. The earth itself is poised between the sun and other planets, and it does not require divine support to float between them. The quantum physicist Carlo Rovelli construes Anaximander to be an early Greek who most helped to lift scientific inquiry off the ground. He and his teacher Thales did not specifically deny the existence of gods; they merely pursued naturalistic explanations that did not invoke them.[15]

Prodicus, living from 460 to 390 BCE, also pursued such a course. As Tim Whitemarsh reports in his impressive book *Battling the Gods*, until recently the work of Prodicus was available to us only through brief reports from other Greek philosophers, usually those critical of him. Then, in a fitting breakout of planetary contingency, a text by Philodemus that had been buried under the massive Mount Vesuvius eruption of 79 CE was recovered, in pretty bad shape.[16] Philodemus himself was a Greek Epicurean who had migrated to Italy.

It was not until a carefully prepared renewal of that text became available in 1996 that richer readings of Prodicus could be pursued. One fragment which Philodemus quotes says: "the gods of popular belief do not exist, nor do they have knowledge."[17] Demeter, the goddess of agricultural bounty, says Prodicus, was actually a woman who had been an exceptional farmer; Dionysus, a talented wine maker; Poseidon, a master seaman. All later fabulated into divinities. Thus, Prodicus begins to naturalize old myths without squeezing out altogether appreciation of periods of nonhuman turbulence in the world from his image of nature.

He supports naturalization of the gods *and* an image of nature at odds with reductive naturalisms and materialisms that became popular in later European thought. The reductive materialists of a later day typically treated human beings as fully susceptible to deterministic explanations, denying them active agency and seeking causal explanations of their conduct. Such formulations, however, face a paradox: Are the explanations by such materialists themselves a product of previous, blind causes or do they rise above the very determinism advanced by the theory? Prodicus does not sink into such a paradox. He seems to subtract gods from the cosmos *and* to retain human agency. He may well have concluded that agency itself emerged as life advanced. Many contemporary biologists can be said to support such a view, too, treating agency to be an evolutionary achievement in a world in which some nonliving processes once bristled with potentials to cross from nonlife to life. Prodicus as naturalist but not a reductive materialist. Did he, however, too strictly restrict striving and agency to humans, losing one insight from Hesiod about a world of multiple agencies even as he subtracted the gods from it? That issue returns.

Another possible difference between Prodicus and later, reductive materialists may be pertinent too. Does Prodicus look forward to a future when the apparent contingencies of nature/culture relations are smoothed out? That is not certain from the materials we have reviewed. Nonetheless, I *extract* from him a view that, since multiple forces and agencies often coalesce and clash, it is wise to assume that nonhuman events will periodically interrupt the trajectories and norms of human culture and experience. A naturalization of the gods joined to preemptive repudiation of later, reductive materialisms oriented to a mechanical world, then, would allow the Prodicus of yesterday to teach us lessons about cultural relations to manifold natural processes of today. It may even help rework the very distinctions between "culture" and "nature" reflected in those last sentences. More about that later.

Of course, as already noted, to the extent they subtracted gods from the cosmos, Prodicus and his ilk courted trouble from Greek peasants on one side and aristocratic Platonists on the other. Even Socrates was compelled to drink hemlock because he was said to have challenged sanctified stories about the gods in his sacrilegious tutelage of young, aristocratic male youths in Athens.

Plato and Augustine, in different keys, will resist with all their discursive might the main trends of thought expressed by Prodicus and others. They will say that Sophists, because they debunk divinely imbued authority, are impious and filled with pride. They are also caught in a series of self-contradictions because they can locate no authoritative, divine source for the public ethos they embrace. Even Marx will later want to subdue the sense of planetary volatility in those Greek stories. And Freud tends to reduce the Hesiod/Homer stories to fabulations of family dramas, out of which the modern need for therapy arises. But in the modern era of climate wreckage—and during a time in which new geological histories point to awesome volatilities periodically punctuating the earth after its inception about 4.55 billion years ago—perhaps more humanistic intellectuals might be wise to pursue trails already opened by Prodicus and others.

Another line of Greek thought treated the gods not only as popular fabulations but also as aristocratic inventions designed (consciously or unconsciously) to infuse piety to the gods and obedience to rulers more securely into the peasant populace. Critias, a relative of Plato and supporter of a violent, successful coup against democratic rule in Athens, is said to have written a (lost) play in which atheism is introduced and attacked as profoundly unpatriotic. In it, Sisyphus is introduced as a partisan of atheistic unrest. He is punished, of course, by being sent to the underworld, where he pushes a heavy rock up the same hill every day, only to have it roll back so that he is compelled perpetually to start again. A perpetual life of futile body building as punishment for impiety. Albert Camus was later to treat Sisyphus as a template for the human condition itself. In a fragment of the Critias play recorded by Philodemus, Sisyphus says:

> Some shrewd man, wise in his counsel,
> Discovered for mortals fear of the gods, so that
> The base should have fear, if even in secret
> They should do or say or think anything.
> So he thereupon introduced religion,
> Namely the idea that there is a deity flourishing with immortal life.[18]

So, Greek doubters and atheists slide between regarding sources of popular belief in the gods as fabulations of extremely talented and socially beneficial humans and the view that rulers and philosophers also promulgate popular piety to the deities in order to support aristocratic rule. Did these two incentives intermingle within a world in which oral reports of heroic, divine struggles accompanied life from the early days of feeding at your mother's breast until death do you part from life? Views imparted to us as youngsters become implanted in both the cruder brain regions (such as the amygdala) of cultural life and more refined thought also connected to those crude regions.

And yet doubts were periodically posed about these gods, too, as we have seen, particularly when new events unsettled life in the *polis*. The gods, perhaps, both signified cultural fragilities that spiral all the way into the heavens and were subjected, in their very multiplicity, to numerous political debates and struggles. They were also potent forces to be mobilized periodically, as aristocrats forged alliances with a section of the peasant populace to overturn democratic rule—at least such as the latter was and could be in Athens. The coup supported by Critias, relative of Plato, exemplifies such a movement.

Ovid and *Metamorphoses*

Ovid was a citizen of Rome who lived from 43 BCE until 18 CE. He was born in Sulma, a small town ninety miles north of Rome; he traveled as a young man to Athens and Sicily to soak up Greek intellectual culture. He was banished by Emperor Augustus from Rome to a small town near the Black Sea in 8 CE, for reasons uncertain. But the banishment may have had to do with his free sexual morality and practices. Augustus, the Roman emperor, lived the life of a philistine as a young man himself, but late in life forbade adultery and abortion. A common enough evolution. Ovid commended both of those latter practices. He advised noble women to pursue adultery, since aristocratic men regularly did. He did not, then, share the overt hostility to women of Hesiod, who considered them a burden for men to bear. Though Ovid could not be called a sexual egalitarian either. The recurrent scenes in *Metamorphoses* where women flee male gods or humans who do or would rape them makes that clear.

We are concerned with *Metamorphoses*, a long epic poem that challenges things in Greek heroic poems but shares with them themes about the origin of the universe and stories about the gods. When you read *Metamorphoses* you note how often Ovid says things such as "or so it is said" when a radical turn has occurred, perhaps intimating to the reader that these stories are less to be taken literally than to be experienced as fabulations that illuminate how the cosmos periodically invades, shapes, and turns human events. In an earlier text, *The Amores*, he says, "the gods, too, if I may be permitted to say so, are made by poetry, and their great majesty needs the poet's mouth."[19]

Book 1 of *Metamorphoses*, "The Shaping of Changes," sets the cosmic stage:

> Before the seas and lands had been created,
> before the sky that covers everything,
> Nature displayed a single aspect only
> throughout the cosmos; Chaos was its name,
> a shapeless, unwrought mass of inert bulk
> and nothing more, with the discordant seeds
> of disconnected elements all heaped
> together in anarchic disarray.[20]

This Big Bang kicks things off. The English word "inert" in the above quote, perhaps, is misleading, for every other word denotes and connotes restless energy, volatile forces bristling with diverse possibilities. Discordance amid metamorphoses, the emergence of new entities out of old encounters.

> The sun as yet did not light up the earth,
> nor did the crescent moon renew her horns,
> nor was the earth suspended in midair,
> balanced by her own weight, nor did the ocean
> extend her arms to the margins of the land.[21]

Ovid seems to accept the Archimedean view that the earth, once formed, took the shape of a sphere suspended in the heavens. He also insists that the sun was formed out of dispersed materials, as were the earth and the moon.

Recent geologists would only modestly rewrite Ovid's formulations. Once the earth was formed and before the seas had taken shape, says Bill McGuire, a contemporary vulcanologist, the moon was probably created after another planet, *Theia,* crashed into the earth, catapulting massive materials from both itself and the earth into the heavens. These fragments later crystallized into the moon, a sphere rotating around the earth in the grip of its gravity. As McGuire writes:

> According to the currently favoured view, however, a very big shock was in store, arguably the greatest in our world's long history. Even before it reached its 100 millionth birthday, it seems likely that the Earth was struck a glancing blow by another celestial body just a little smaller than Mars. This object, sometimes going by the name of Theia—the Titan in Greek mythology who gave birth to Selene, the Moon goddess—is charged with gouging out a sizeable chunk of our planet's crust and mantle . . . [that went on,] along with some remains of Theia, to form the Moon.[22]

The result may have shifted the tilt of the earth; certainly, after the oceans were formed later, the moon did create tides and a guide for sailors at night. Ovid, I suspect, would be pleased to accept this refinement of his story. Metamorphosis.

Ovid's genealogy of the earth continues.

> Although the land and sea and air were [now] present,
> land was unstable, the sea unfit for swimming, . . .
> and all things were at odds with one another.[23]

It took "some god" to settle "this dispute": it "disentangled" these elements, separating earth, sea, sky, and heaven. And now the god, "whichever one it was," could mold the earth "into the shape of an enormous globe."[24] So, it now gives shape and solidity to the entire planet. Note how Ovid saw the earth as a globe well before Christians did.

Now "man was born." "Man," however, goes through four stages, each worse than the one preceding it, roughly following the pattern described in greater detail by Hesiod in *Days and Nights.* Eventually the degraded humans demand too much from the soil, and persistent feuds between them prevail. And Ovid is alert to how deforestation to increase agricultural production also reduces the fertility of the soil and can lead to desertification.

Eventually, Jove, the Roman handle for Zeus, decides the human race must be destroyed so it can start again with a new slate. Some of the gods, however, give only "mute assent" to this proposed holocaust, perhaps signaling cautiously their reservations. A Great Flood is enacted.

Now with his trident, Neptune strikes the earth,
who shudders at the blow and opens wide
new waterways.

Soon,

There are no longer boundaries between
earth and the sea, for everything is sea,
and the sea is everywhere without a shore.[25]

Jove is pleased to see that only one man and woman survive. They are the two "blameless and devout" ones.[26] They are assigned to couple and start the human race again.

The story, of course, resonates with that of Noah and the Ark in Hebrew and Christian traditions. And, indeed, there is recent geological evidence that a huge flood did ravage the Mediterranean world about 7,000 years ago when the Bosporus straits were flooded massively by a rapid rise in the Mediterranean Sea, producing the rapid expansion of a large lake bordered by farms into the Black Sea.[27]

Did Ovid and the Israelites both fabulate (in rather different ways) a horrific event that actually occurred, one causing death to many farmers on the shore of a smaller lake on the edge of modern Turkey, forcing innumerable others to migrate south and yet others to head west, a gruesome, unsettling event that seemed to threaten human life everywhere? Was this event then amplified and moralized differently in several cultures to help shape divergent religious traditions as it was passed down orally to descendants over the centuries? The contemporary evidence is suggestive but incomplete. It is interesting, though, that Ovid was banished to a small town near the shore of that great body of water now called the Black Sea.

In the Ovid version of the story, the man who survives is distressed about the horror the god has wrought; he asks his wife how she could have borne this terror without him or he without her.

"Who would console you in your unshared grief?
For trust me, if the sea had taken you,
I would have followed then."[28]

The lone survivors thus do not treat the holocaust as if it were deserved. They seem to regard the gods as rocky, volatile, personified forces to be taken cautiously into account and sometimes placated, more than as just and benevolent beings fully worthy of pious worship and obedience. They reduce, without eliminating, the drive to self-castigation in the human relation to the gods.

Well, in this new world the stories of transformation continue. Men are transformed into women, women into trees, horses into humans, humans into gods, maidens into warriors, weavers into spiders, and so on, all under the canopy of bumpy gods/human relations. Nothing, neither thing nor being, is treated as if it is permanent; and the rapid pace of transformational moments can make one dizzy. It feels as if the stories themselves are organized through dramatic compressions of time. Life is a riot. And time is bumpy.

The interventions of the gods compress earthly time. To many of us today, it does seem plausible that life emerged out of nonlife as complex molecules of diverse sorts combined through chancy conjunctions to generate life and as life itself then began to take on a wondrous heterogeneity of forms. But at a slower pace. We are also impressed with the metamorphoses that sometimes occur when the entry of a sperm or virus into an embryo initiates resonances that render the new birth different from its two sources. Metamorphoses. In Ovid's stories transformations of different sorts often appear suddenly, thanks to interventions from the gods. These restless gods are temporal accelerators. Consider a few examples.

Juno, previously known as Hera in Greece, found the sexual escapades of her brother/husband, Jove, to be intolerable. She was overwhelmed by divine jealousy. After Jove took Semele, a beautiful woman, as his lover, and Dionysus was born of their union, Juno had had enough. "Why is she a mother of your child and not me, a pleasure I am denied," Juno asks. So, she had the child torn out of Semele's womb before it was ready to be born. But some nymphs intervened, "hid him in their cave / and nourished him with milk."[29] Thus, Bacchus (Dionysus in the Greek myth), is "twice born." He, conceived through a union between a mortal and the most powerful god of them all, becomes something of a rebel among the gods.

Meanwhile, the issue of who gets the most pleasure in sex, man or woman, becomes an object of intense debate between Juno and Jove. When the question is posed to her, Juno lies to Jove, saying it is certainly men. Tiresias is a "transsexual" human, a term that in this context seems to mean one who takes erotic pleasure with both men and women and perhaps also combines the sensitivities often attributed to women with a fierceness often attributed to men. So, Tiresias was consulted by Jove to check out the veracity of Juno's story. Tiresias said, "women get far more pleasure out of sex than men." Did the transsexual, in a patriarchal hierarchy of gods, men, women, and slaves, try to suggest to Zeus that this secret was kept from men so that women would not have their own passions, too, used against them in a patriarchal order?

We don't know. But Juno reacts to that insult by striking Tiresias blind. Jove soon took pity on the fate of Tiresias; but even he could not directly undo what another god had done. He was powerful but not omnipotent. Circumvention was the route to take. So "Jove gave Tiresias the gift/of foresight to replace the vision lost."[30]

Tiresias now becomes a blind seer, with the capacity to look into the future with insight but not to attain exact knowledge of it, the very seer consulted in Thebes when tragedy was at hand in two plays by Sophocles, *Oedipus* and *Antigone*. For the early books of *Metamorphoses* return to—and often rewrite in this or that way—orally transmitted stories Homer and Hesiod had transcribed in Greece.

Let the metamorphoses of Tiresias stand in for myriad other compressed transformations, often presented with humor and a degree of textual distance from the gods said to spawn them. The gods compete with each other during many of these transformations. Humans, exercising judgment and agency of their own, are often caught in divine turmoil. Tiresias is punished when he innocently reports his judgment about the relative sexual satisfaction of the two sexes when a god posed the question to him . . . He is the target of divine conflict in the skies. Juno later punishes Semele, too, after Jove had shared a bed with her. We do not know whether Semele had consented to that union with the most powerful god. Rape by male gods of fleeing women and goddesses is more than amply recorded in the *Metamorphoses*.

Later in the text, Aegina, an island off the Greek coast with a regime sometimes in conflict with that of Athens, is struck by a devastating plague. It is reported by Ovid—of course—to be brought on by the "cruel Juno."[31] She inserts the disease in a way rather close to how we think a virus does. His description of the suffering is vivid. We pick up the story after the plague has crossed from other animals—"birds, sheep, cattle, and wild beasts"—to humans. Ovid was an early believer in species disease crossings. As Aeacus, king of Aegina, relates:

> "The plague, grown stronger, now advances on
> the wretched country folk, then rules within
> the walls of the great city. Its first symptom
> is a fierce burning in the viscera."[32]

Soon doctors themselves begin to die. And it spreads to the entire populace.

> "Some freed themselves from the fear they had of death
> by taking their own lives—summoning Fate
> even as Fate prepared to summon them."[33]

Ovid is of course aware of plagues that had struck other Greek city-states, including Athens. Another, the Antonine plague, was to strike Rome itself in

165 CE, probably contracted by Roman soldiers during a campaign away from Rome and carried home. The Justinian plague struck Rome later, during the years 541–49 CE; it combined with the "Antique Little Ice Age," nomadic invasions, and internal conflicts to help pull the empire down.[34]

Ovid, as we shall see, does not anticipate such an outcome for Rome. He lived in a period of Roman expansion during what is now called the Roman Climate Optimum. Nor did he have Tiresias to consult. But he is nonetheless often alert to the gods' infusion of poisons into humans that contributed to making life vulnerable and precarious for towns, cities, and the countryside.

Metamorphoses, with its transformations, simultaneously entertains and disturbs with its manifold cruelties attributed to gods and humans; it also illuminates the bumpiness of life by invoking gods as personifications who insert periodic tumult into the human world from the outside. The text does not seem, however, to demand that everyone take those divine sources of dramatic turns literally. Only that there are such turns, some set on a fast scale of time in *Metamorphoses* to dramatize them to adults who can themselves expect only to live forty or so years if they survive the perils of infancy. (The average life expectancy was twenty-five years, but that figure includes the large number of children who died young.)

One function of those invocations of gods, again, is to compress temporally some events that would otherwise be held to take longer. The gods thus alert us to how we so often act as if things that have persisted for only decades or centuries—say, a climate pattern, or a democracy, or a pandemic-free period, or an ocean current, or capitalism, or a period of peace—carry an illusory appearance of permanence.

Take the story of Echo and Narcissus. Echo, the beautiful nymph who has not yet lost her body to become a mere echo, is madly in love with the impossibly handsome Narcissus.

> Narcissus at sixteen seemed to be both
> boy and man, and many boys and women
> desired him; but in his yielding beauty
> was such inflexibility and pride
> that no young man or woman ever moved him.[35]

The story of his life is condensed into a short period of weeks. He is madly in love with himself—seeking urgently to close the distance between himself and the beautiful image of it he observes in a limpid pool.

Echo, who had been condemned by Juno for uncertain reasons not to speak in whole sentences but only to repeat the tail end of phrases uttered by others, becomes entranced with Narcissus. She "grew hot" at his very sight;

> she secretly pursued him through the woods,
> her heat increasing as she overtook him,
> as torches smeared with highly flammable
> sulfur ignite themselves, brought near a flame.[36]

But she herself cannot initiate phrases needed to draw him to her; she can only echo fragments of sentences uttered by him. As she approaches, Narcissus asks if anyone is here. "*Here!*" the echo responds. She repeatedly approaches, and he evades. "I'll die before you have your way with me," he finally says, and Echo resounds, "*You'll have your way with me.*"[37]

Narcissus spurns her. She is soon reduced to an echo without a body, able to amplify the closing words of others but not to inhabit a body, to mingle with other bodies, or to speak full sentences in her own voice. The lovely Echo, whose big mistake was to be too hot for a narcissist, is doomed only to echo the voices of others, reminding them and us, as she does, how often we hurl words into a void that overwhelms them. The sad being now personifies those echoes we hear in the mountains, reverberating echoes that also remind us of the vast size and range of those peaks.

Today we increasingly inhabit a world populated by echoes without habitats. The pace of climate heating disrupts rhythms; species arrive at old haunts to reproduce, only to find their timing to be out of sync with the setting. Take the threatened songbird, iguaca, in Puerto Rico, whose habitat has been upended by more intense hurricane seasons. Recordings of its songs now ring hollow; they echo lost worlds.

Narcissus himself soon stares at his visage in a clear pool. He falls even more madly in love, seeking to merge with it. Leaning over, he falls into the pool, shattering the image with which he sought to merge. He dies a sad death, disconnected from the only thing—his own image—he could love.

The story of Echo and Narcissus is compressed into a short compass. But, in fact, the vicissitudes of narcissism often take an entire lifetime to unfold. Temporal compressions through the instrument of divine intervention perhaps allow us to ponder such lessons more reflectively, as interspecies and earth rhythms increasingly fall out of sync with each other. The story also provides a template to the dangers of narcissism in the academy.

Can it teach intellectuals how the pursuit of cultural self-consciousness through introspection—that is, by peering into the depths of one's self—can readily collapse into narcissism? Does it even teach that introspection is an insufficient way to deepen or move the sediments of one's own thought? Descartes, for example, will later advise intellectuals to doubt everything until, through rigorous introspection, they reach that which cannot be doubted: "I think therefore I am." He concludes that on that slender base you can build a whole, certain philosophy. But is perhaps what is taken to be indubitable through peering into your own soul less a *test* of your most basic ideas and more a *reflection* of that which in your culture and personae has become difficult to think otherwise?

Is Cartesianism itself a mode of collective, human narcissism, pulling us further away from the multiplicity of life forms that inhabit the earth rather than toward closer communication with them? Perhaps *comparative* analyses of the depth grammar of our own society, philosophy, or theology in relation to other human and nonhuman orders is a better way to pursue the agenda of deepening cultural reflection. The introspective method can even lead to what Michel Foucault called "transcendental narcissism"; it can be defined as the tendency to freeze, as if they were indubitable and undeniable, cultural postulates upon which a particular regime of thought and being has been founded: of the gods, an omnipotent god, the self as unified subject, or humans as the only reflective modes of life on earth.

Of course, you can't compare your own world to all other possibilities in adopting such a method of comparative genealogy. Life is too short for that; prescient choices must be made. But several modern western thinkers, Nietzsche and Whitehead among them, contend that the "searchlight" of self-consciousness through internal perception—or the self-reflection beloved in different ways by Narcissus and Descartes—is weak. Supplementary methods of working upon one's own intellectual assumptions are needed.

Ovid agreed. Perception is outward oriented; it is less effective when trained into an internal spotlight. You may think you are peering deeply into your self when in fact you are wiring more deeply into yourself contingent features of life already branded into a regime, a theology, or a philosophy.

Thinkers such as Nietzsche, Foucault, and Whitehead, after advancing their own critiques of transcendental narcissism, definitely don't counsel sinking into tradition uncritically. They, rather, revise and rework the dialectic of self-consciousness that has influenced modern western thinkers at least from Descartes, through Kant to Hegel and beyond. Nietzsche, for instance, deploys "genealogy" to suggest how a variety of unities—a God, a cultural morality, an image of self, clear and distinct ideas—all taken at different times to provide the stable foundations and measure of life, are actually formed from disparate elements through a series of disciplinary processes. I wonder whether Ovid was an early advocate of such a double view: a critic of introspection and a devotee of alternative practices of testing consciousness. Or do I read too much into these stories?

One theme of *Metamorphoses*, as of the *Theogony*, is that turbulence in the world is often not the product of gods or humans alone. For passions circulate between gods and humans, often becoming amplified on both sides as they do, heating up conflicts and setting into motion new transformations. The ways of the world thus involve multiple concatenations between what many in modern life, at least until recently, would call "nature" and "culture," as if the two could be neatly bifurcated. They cannot. Ovid knows otherwise.

Moderns often used to say the Greeks and the Romans conflated nature and culture. But the tides are turning. Think about how accumulated capitalist fossil emissions heat up the atmosphere, setting into play, as they do, multiple planetary amplifiers that soon roll back to haunt the world with greater intensity. These impersonal, nonhuman amplifiers, then, distribute some of the worst effects to regions populated by people of color who have made the least contribution to the triggers that set them off.[38] Impersonal circuits of imperial oppression added to those intentional ones already in place.

Let's consider just one more story in *Metamorphoses* before turning to its last chapter. The young maiden Arachne became famous as a weaver of incredible skill and speed who wove fantastic patterns. The supple, resilient yarn could assume multiple shapes and patterns, and the weaver was less its master than one who listened to its protean possibilities to cooperate with them. Almost like a spider tends its web and the web tends the spider, whose small brain is amplified by the extensions it weaves. The focus on weaving in this story may resonate with our later engagement with how it is said in Aztec culture to provide the means by which the cosmos itself is held together.

But Minerva (Athena in the Greek version), a weaver herself besides being a goddess of wisdom and war, became insanely jealous of this skillful, sensitive earthling. Disguising herself as an old woman, she challenges Arachne to a contest. A weaving competition is arranged:

> Then they go to it,
> hitching their robes up underneath their breasts,
> their well-instructed fingers swiftly flying,
> and zeal for the contest making light of labor.
> Into their fabrics they weave purple threads
> of Tyrian dye, and place beside them shades
> that lighten imperceptibly from these;
> as when a storm ends and the sun comes out,
> a rainbow's arc illuminates the sky;
> although a thousand colors shine it,
> the eye cannot say where one color ends
> and another starts, so gradual the verging.[39]

Arachne eventually wins this cosmic contest. As her patterns unfold they portray, among other things, the cruelties of the male gods to women they so often raped and always subordinated. Minerva, the powerful, jealous goddess, and favorite of Jupiter, decides to punish her. Minerva is, after all, a daddy's girl, born from the forehead of Jupiter in order to evade the participation of a pregnant mother.

> Then, as the goddess turned to go, she sprinkled
> Arachne with the juice of Hecate's herb,
> and at the touch of that grim preparation,
> she lost her hair, then lost her nose and ears;
> and her head got smaller and her body, too;
> her slender fingers were now legs that dangled
> close to her sides, now she was very small,
> but what remained of her turned into belly,
> from which she now continually spins
> a thread, and as a spider, carries on
> the art of weaving as she used to do.[40]

The spider shape and striving is now inhabited by a subdued human, encouraging it to strive to become other than it is. It may encourage you to think about spiders differently. When a spider weaves its web, it is now known, flies who seek to avoid it by merely flying close are making a big mistake. The web secretes a magnetic field that pulls them into its trap.

Perhaps one way to engage *Metamorphoses* during a time of viral crossings, fast evolutionary changes, and climate wreckage is to think of it as Ovid's deployment of the gods to *compress* radically the multiple temporalities or vectors of different beings and forces that together constitute the messiness of time. Not Augustinian or Cartesian or Kantian or Einsteinian images in which time is set on a singular course or governed by one measure. But a messier time, composed of multiple trajectories of different beings and forces, periodically overspilling into one another. Or colliding. Or resonating together, as a violin and cello do in a musical score. Out of such intersections and resonances new entities come into being, themselves set on new courses soon to converge, collude, or collide with yet others. You yourself are a few vectors, my friend. So are the civilization you inhabit, the volcanic eruptions that turn it, the election cycles that weave into it, and the other civilizations upon which it feeds. Viruses, planets, bacteria, ocean current systems, election cycles, planetary orbits, atmospheres, beehives, and whales ride such currents, soon to collide with others. Think of that spider you note crouched in its web on the garage door as an arachnid, composed of multiple strivings, bodily contours, a web, and the atmosphere it breathes, perhaps to bite you, or to inspire you to think in new ways about webs you inhabit, or to watch it capture a fly in its sticky web, or to note how it provides silk for the shirt you wear. We can better surmise what it does and how it secretes its silky threads, perhaps, if we, too, on occasion fabulate it to house the soul of the great weaver, Arachne. Punished by a jealous goddess. And don't forget that life itself as we know it today is rendered possible by the oxygen that phytoplankton, trees, and plants release as waste. The planetary self-oganization of photosynthesis provided quite a turn of the earth, a metamorphosis within life.

Arachne, weaving with fine, supple wools, and the spider, an arachnid weaving webs with silken fibers rich in resilience, suppleness, sensitivity, and outreach, so that a slight bounce anywhere in the web alerts the spider to pounce as the fibers delicately slide back into their previous form. These relational images of weaver and fiber, this grasp of how weave, viscous materials, and weaver are intertwined, did they not help to sow the seeds of future sciences of complexity? Were those latter sciences, which emphasize the fecundity of relations between diverse entities, unnecessarily delayed in coming to fruition by the combined forces of compulsory Christianity and secular sciences anchored in a world of solids?

Ovid might suspect so. So do Michel Serres, Laura Tripaldi, and Jane Bennett.[41] In these latter views, intelligence, intention, feeling, memory, and sensitivity are not confined to one of two worlds, of mind rather than body, or even, in the extreme, of humans rather than nonhumans. On the cusps between those crude categorical distinctions reside viral crossings, bacterial strivings, slime mold collectivities, spider silk capacities, horizontal gene transference, plant intelligence, and fungal networks, all exhibiting modes of intelligence, memory, and sensitivity that belie the crude images through which moderns have approached or ignored them. Even the octopus, preceding humans in evolutionary time by millions of years and operating on a parallel line of evolution, displays effective modes of intelligence and responsiveness very different from ours. It coordinates a small brain in its head with eight in its arms.

By speaking of compulsory Christianity I do not project forward a world without Christianity. The phrase is meant to resist demands, operative for centuries and still in play, that no other cosmic orientations be allowed to flourish in the same territory alongside it. What if instead of compulsory Christianity officially taking over the terrain of myth in western Europe for several centuries, the myths of Ovid had persisted alongside it with a similar standing, challenging its claims to hegemony on the same territory? Would the late-modern sciences of complexity—sciences built around a multiplicity of agencies and relational fecundities—have arrived sooner? Would biological theories of evolution have become more bush-like earlier rather than being confined to their official tree-like model for so long? Would ecological cultures have thrived earlier and periods of reciprocal exchange between indigenous and Christian/secular cultures have thrived more? Such counterfactuals are huge, but facing what we do today, it may be wise to entertain some of them nonetheless. They help to free the intellectual imagination during a time that was not expected by most western geologists, humanists, Christians, and secularists before the 1980s.

Ovid, if you accept his favorable references to Pythagoras on the issue, supported reincarnation, as have many Hindu, Buddhist, and Amerindian peoples. And, indeed, the stories he tells show humans becoming incarnated into insects, trees, birds, and the like. It is illuminating, initially, to compare reincarnation to the competing, exceptionalist Christian practice of burial in iron caskets bound to faith that humans are uniquely eligible for heaven or hell after an earthly death. Those devoted to Christian orthodoxy demand closed iron caskets to protect them from being digested by insects, bacteria, wolves, or other beings that lack souls. The vision of reincarnation, by contrast, counsels us to respect the agencies and capacities of multiple organisms of heterogeneous sorts, some of which may have human souls trapped in them.

Pythagoras tried to respect all animals by eating only vegetables. But what if, as Ovid himself could have surmised and as an increasing number of biologists suggest today, a variety of plants, too, express differential capacities to feel, to think, to decide, and to suffer?[42] Would that now mean a caring vegetarian can no longer eat anything?

An alternative way to think about this is to say that while multiple species find a variety of others to be delectable—as the crocodile, the wolf, the vulture, the shark, the cougar, the tiger, and microbes do human beings—the highest human response to the general condition of *edibility* is to treat all living beings with respect even as we eat some of them to nourish ourselves. Even as they eat us, if the need and opportunity arise, in a world of general edibility. You would now oppose those factory farms that reduce pigs, chickens, tomatoes, and cows to mere commodities. And you might refuse to be buried in a metal casket, planning instead a green burial to enact appreciation of how profoundly connected humans are to other modes of plant, bacterial, and animal life. Embracing a world in which all living beings become compost for one another. Such a response is closer, perhaps, to reflections by Val Plumwood and Ovid than to either a Christian idea of unique human immortality or a secular treatment of nonhuman flesh and plants as mere commodities, ready at hand for human exploitation.[43] Particularly when you bear in mind that Ovid often seemed to treat those gods who presided over metamorphoses as figures more than beings.

In a world replete with tragic possibilities, there are few pure solutions to every issue, but some responses may help more of us to infuse appreciation of the multiplicity of living beings into how life is lived. The thoughts suggested above would already require massive civilizational changes to be enacted. And that is just the start during a period of climate wreckage.

Ovid, remember, was later demeaned as a pagan by Christians and by many secularists who followed them.

By the time you reach the fifteenth chapter of *Metamorphoses*, the theme of the changeability of beings and things has been sufficiently pressed upon and infused into you. Ovid there offers an appreciative reading of the Greek sage Pythagoras, who eventually settled in a town on the heel of the peninsula now called Italy. Ovid deploys this dream figure to remind us that

> "Nothing persists without changing its outward appearance,
> for Nature is always engaged in acts of renewal,
> creating new forms everywhere out of the old ones."[44]

Now, at least on my reading of that chapter—titled by the translator "Prophetic Acts and Visionary Dreams"—the epic takes another turn. Julius Caesar, who had recently been killed by those protesting his autocracy, is dreamed up into a night star, to shine over the Roman Empire permanently. And Augustus, the new emperor who is about to banish Ovid from Rome to a cold outpost of the empire, is similarly deified. It seems the Roman gods are now defined as providential, guiding the leaders of Rome and protecting the empire. Apollo serves as patron saint of the empire.

Did Ovid take this turn out of fear of being banished by the emperor? Or was it, rather, a strain in his thinking all along? Perhaps it was the latter. But if it was, other strains were in play too. He could have, for instance, followed one to emphasize how fragile and susceptible to external and internal disruptions Rome itself was. He could have emphasized how nothing is permanent, not even Rome or the star to which he catapulted Caesar. But he did not. Later Pythagoreans, such as Ocellus Lucan, writing shortly after the death of Aristotle and Iamblichus (250–325 CE), insisted that Pythagoras, the prophet of resonances between musical harmonies and a harmonious cosmos, had anchored his philosophy in confidence that the cosmos itself is eternal in form. That cosmology fits better, I think, with Pythagoras than one, such as Ovid's *may* be, that begins in chaos and heads toward harmony.[45]

At this point, I as a theorist living during a period of climate wreckage and striving to learn from Ovid find myself pulled to work upon the text from which I draw sustenance. I think this almost always becomes the case at some points in textual interpretation; it is, to me, better to acknowledge how and when you find yourself *working upon* a text from which you draw sustenance than to pretend your preferred outcome was always "implicit" in it all along. You move to and fro, working upon it as you allow it to work upon you. Trying to keep alert to problematic elements in the interpretation you are advancing as you advance it. Keeping one eye on Ovid and another on how it might illuminate surprising events in the present. That is what I will begin to do in this last section.

Ovid has often been judged to be callous and indifferent to human suffering, even to be degenerate. One book about later receptions of Ovid in Europe emphasizes how that theme has recurred over the ages.[46] Perhaps he was. Another way to try to read him, however, is to say that he commends double-folded readings of events in the world. For a time, you view the play of events as if from the top of a mountain, observing innumerable small creatures bouncing around below you. Such a perspective may help you come to terms with comic and tragic potentialities in life itself. Next you move the camera in closer to the Ovid scenes, striving to appreciate the strivings, achievements, conflicts, and suffering more intimately. You thus allow a *double vision* to help you come to terms with the world, as you explore intelligent ways to respond to it.

Is this a sufficient way to interpret Ovid? Probably not. But he does offer a close-up of that plague in Aegina, and the fates of Echo and Arachne also seem to capture his concern. Even if he does not himself adopt a double vision often, however, it may be a thoughtful way for others to *respond* to his stories, to fold them into an ethico-political orientation partly inspired by him. You conclude that many achievements in this world are more fragile and precarious than participants immured in them often acknowledge, while you also work hard to protect and improve through revision some of the most important among them. Such a double set of responses looks more promising if you have already decided, say, that Kantian and neo-Kantian singular orientations to the world too often paper over their own cruelties in the name of human goodness. But we have yet to test that latter judgment. We will do that later.

Providence, Fate, Contingency

Consider two sets of contending terms, with each item in each set having affinities to others in its set and contrasting to those in the other set. Crossings, however, sometimes occur between sets, because the items in each remain soft in the middle and fuzzy around the edges. These are protean crystallizations, open to new adjustments.

The first set: providence, divine purpose, divine guidance through piety, freedom as realization of cosmic guidance, eminent cause, and fate. The second: fortune, contingency, chance, event, emergence, and—again—fate. To paint a Greek face on the contrasts between the two sets you might label them to reflect differences between Apollonian and Dionysian views of the world. How sharp are the differences between the two sets? Which set provides the most insightful premonitions from which to draw? What distinctive sense does the notion of fate assume in each set? Why are some of us instinctively drawn to the first set while others are pulled toward the second? Is it best to pursue a balance between the two sets or to pick one and seek to make the most of its complexities?

You might say, for starters, that in the first set fate means, roughly, the capacity of a single god or the gods to shape key outcomes in the world. The doctrine of predestination in Augustine and more radically in Calvin, takes that form, too. On such a reading, the fate of Oedipus was foreordained by the gods before he was born.

According to the second reading, however, fate means something closer to a set of diverse temporalities or trajectories that come together in an unexpected way to set a new condition facing people and other species too. Fate is not preordained, then, but when it congeals it sets a new context as old priorities lose their grip and new ventures proceed. People who embrace fate in the first sense, then, are said by those adopting the second view to invert things, to protect the gods by regularly insisting that the relevant divinities intended before the event what happened during it. The latter perspective on fate, you might say, unfolds if some themes from Prodicus are poured into Hesiod and those of Hesiod into Prodicus. A world marked by periodic planetary volatilities that help to shape fate.

Very few naturalists, during this long period in Greek and Roman thought, accepted anything close to a mechanistic notion of causality; so they would not have defined "chance" to be a mere break in an otherwise tight causal *mechanism*, similar to a car breakdown. The modern determinism/freedom debate was not really on the agenda for Ovid, since bodies were not then conceived as mere solid mechanisms. But debates between preordainment and uncertain torsions between contingency and intersecting agencies were very much in play.

Perhaps the focus for now, then, should be on the conflict between one view in which either divine providence or an ugly divine fate plays the defining role in life and another in which fortune or contingency periodically plays a surprising role. When a new threshold is reached and the gods are favorably disposed—clearly, that is not so often the case, as we have seen—divine providence becomes important to thinkers such as Hesiod and Plato. But what if you contend, with atheists of that day, that there are no gods; or with the

Epicureans of the day, that "thin" gods subsist in the stratosphere but do not interfere in human affairs; or with Ovid, that the gods are fascinating figures through which to record and compress temporal transformations but perhaps not real beings?

Here, of course, effective human leadership now becomes important to forestall the worst, as does the task of convincing a diverse populace to accept your wisdom. But contingencies periodically intervene to challenge, redirect, or overwhelm those capacities. Contingencies are events that, even in cases where they are potentially susceptible to explanation in the future, take a constituency or a regime by surprise. They thus rattle established patterns and contemporary projections into the future. The big ones even seem to turn time. Ovid's Great Flood would be such a planetary contingency to Prodicus, as would the eruption of a plague in Athens, or the capture in 1619 of Congolese by the Portuguese to initiate a new slave trade, or the rapid worldwide spread of COVID-19 in 2020. In some of those events, several actors act with intention—take the intentions and goals of the Portuguese invaders, for example—but the receiving population encounters the intervention as a surprise that profoundly interrupts conscious and unconscious extrapolations they have heretofore made into the future. It might in principle be possible today to anticipate a new pandemic; however, the likelihood that viral experts with power to communicate internationally will always be in the right place at the right moment is very low. Hence, pandemics will often be contingent events.

On this view, specific social trajectories are periodically interrupted by events in a world not reliably or always predisposed to human welfare. The events often disrupt common extrapolations into the future built into the institutions and popular stories of the day. The same event can encourage some to change their projections and others to double down, to cling all the more obstinately to old projections now under threat. To strive to fend off a new future by holding more firmly to an established image of transcendence. Reactions to the failure to attain a second coming, to rapid climate wreckage, to COVID-19, and to the 2020 American election of Joseph Biden all express such diversities of response.

Such an evental image of the role of fortune and contingency does not deny a role for purpose or striving, as some versions of modern, reductive materialism do. Rather—in a way that draws it a bit closer to a world of multiple gods—it projects a world populated by multiple strivings, purposes, and forces of numerous sorts. To put it in modern language, clashes might periodically occur between the nano-strivings of viruses and human tissue, or between bacteria and humans, or between the acceleration of climate change and the vicissitudes of politics in different regions of the world, or between a rapid mountain glacier melt and the dependence of animals, plants, and humans in the valley below it upon slow, cyclical melts, or between a fish with large fins, new sources of food on land, and the metamorphoses of some sea animals into those with limbs who walk on land.

On such a view, not every entity in the world searches and strives, though multiple entities of widely diverse types do. You can include humans, bacteria, wolves, slime molds, spiderwebs with magnetic powers, sunflowers, vultures, Venus flytraps, jaguars, viruses, mosquitoes, octopi, fungi, and whales in that latter set, in both their individual and collective modes. Others, though, are blind forces, even if they do acquire acute *sensitivities* to unusual perturbations on occasion, sensitivities that may propel them in surprising directions or move at unexpected speeds. You might include volcanoes, earthquakes, climate amplifiers, methane, glacier flows, trade winds, and ocean currents in the second set. They, too, periodically coalesce or clash with each other *and* with the first set of trajectories. Such conjunctions can help to create preconditions for a whole civilization to flourish—as that three-hundred-year period of Climate Optimum did in early Rome. Or they can contribute to civilizational fragility, as conjunctions between pandemics, the Antique Little Ice Age, internal conflicts, and external invasions did later in the same Rome.

Some thinkers, such as Pythagoras, Plato, and Parmenides in different ways, emphasized the role of piety and divine providence in promoting harmony in the world, even sometimes arguing that noble lies are needed to imbue the populace with needed pieties and dispositions to obedience. They were carriers of what might be called the idea of eminent causality, *the idea that something complex can only find its source in something of yet greater complexity.*

Their competitors, such as Anaximander, Lucretius, Epicurus, and Prodicus, suggest that such premonitions of providence misread the fundamental contours and trajectories of the world; they underplay potential clashes and assemblages between diverse human strivings and multiple other agencies and forces of the world. They also place too much trust in leaders endowed with the authority to interpret the veiled will of god or the gods. The civic virtue lauded by teachers of divine providence, these critics argue, too readily becomes mass obedience to divinely inspired rulers.

The critics of eminent causality often become proponents of *emergent causality*, the idea that diverse and sometimes complex modes of being periodically emerge out of bumpy intersections and sneaky insinuations between less complex processes. Evolution is founded on such a principle.

For reasons that will progressively become filled out, I feel closer to the second set of orientations than to the first. Why, then, pay attention to Hesiod and Ovid? Why not merely skip them? Well, I may break with them around some questions about the gods, authority, piety, and obedience. But they emphasize, more than do several other thinkers of today and yesterday, the turbulent origins of the world we inhabit, its periodic unruliness, and the periodic emergence of new forms out of old coalescences.

Hesiod and Ovid thus help to keep modern, western thought on its toes, speaking to it across a temporal chasm of centuries. They make us think again about how large planetary forces impinge upon the cultural lives of humans and other species, how it is wise to fold such infusions and impingements—often surprising—into thought about the world itself, and how to struggle so that such intrusions do not become pegs for popular autocratic drives.

It is perhaps easy to see how people struggling amid numerous insecurities, who have placed their trust in gods or a god, could go berserk when events and social movements call some of those very modes of piety into question. That may pose one of the dilemmas of late-modern life: to fail to contest some faiths over the human capacity to control nature and/or the prominence of divine providence is to court danger; to do so is to court danger, too, from different places and sectors.

Does that mean, then, that to contest equations between providence and piety it becomes necessary to become detached from the world? Certainly, several modern secular theories have taken such a turn. But it is not necessary to go that way. From the point of view now on the table, such a positivistic turn is also mistaken. The cultural metamorphosis of tempestuous gods into a host of intersecting forces and agencies, rather, can encourage more of us both to become more attuned to periodic volatilities of the planet and to deepen (sometimes nontheistic) reverence for its grandeur. Hesiod, Prodicus, and Ovid offer tantalizing leads in this respect, as do others yet to be consulted.

Periodic turbulence is part of the earth itself on this account. And that very world also enables us to be. Some of its fluidities and multiplicities inhabit us; others surround us. It challenges us to act creatively from time to time in response to new events. And, when we cast off the authoritative piety and obedience to singular rulers pushed by the first orientation, we may become eager to explore new, democratic modes of collective attachment. We might even introduce nontheistic reverence for the world as an active existential creed through which to respectfully contest *both* providential and detached visions. We might project, that is, strains of deep pluralism into modern cultures as we refuse to sanction autocratic eliminations of them. And we might extend vaunted human notions of agency and striving onto a larger variety of nonhuman processes, many of which live within us and others of which act upon us from outside.

Much of modern secular theory, of course, grounds itself neither in the pieties of Hesiod to the gods nor in the tumultuous stories of Hesiod and Ovid.

On the other hand, such modern capitalist, socialist, and democratic theories have too often been grounded in the presumption that "nature" can be our oyster, *either* to master *or* to become more attuned to as a fount of stable harmonies. Either/or. The debate between these latter two visions has often prevailed in modern life, with the first striking the major chord and the second the minor. But *neither* projection has really worked out that well.

The view anticipated in these comments, then, contests two, preset, modern alternatives. It calls into question the sufficiency of the debates between them. Doing so, it also appreciates a strain discernible in both Hesiod and Ovid. The forces of the larger world—either defined as forces and agencies or fabulated into multiple gods—are not automatically predisposed to human flourishing. Such a world, indeed, might excite our appreciation of its grandeur as we also struggle against the dangers and suffering it periodically spawns. The cultural assumptions made by many Euro-American thinkers during the modern age that the world is either susceptible to harmony with us *or* to our mastery can readily turn into patterns of profound disappointment or nihilism when events rattle the terms of that debate. And events keep coming. How to appreciate the volatility of the world without lapsing into nihilism as old debates between providence and mastery become unsettled? That is the question.

A further thought. It does seem plausible to say that in both Hesiod and Ovid the planet begins in turbulence and, over time, wears down into a cyclical order in which the edges are worn off those turbulences and volatilities. Things descend into greater order, replete with cyclical variations. Given such a development, can these texts really be drawn upon to incite more creative thinking about today, the time of climate wreckage—a time marked by the renewal of wildfires, intense hurricanes, more volcanoes, atmospheric rivers, enlarged algae blooms, and release of clathrates into polar zones? Maybe these two texts can be *worked upon* modestly, to show how old volatilities become ignited in new ways under the pressure of new events. No tradition is perhaps by itself fully sufficient to the time of climate wreckage. But some may be more suggestive and replete with fecundity than others.

A related thought is worth noting too. In *The Fifth Hammer: Pythagoras and the Disharmony of the World,* Daniel Heller-Roazen exposes how Pythagoreans explored, through analogies between musical harmonies and mathematics, the vision of a world of harmonies. But they constantly ran into dissonances that troubled their aspirations. One story, about a Pythagorean who was killed because he publicized such dissonances, is probably apocryphal. But it is revealing nonetheless. Heller-Roazen explores such stymied Pythagorean attempts all the way to Kepler in the 1600s.[47] Suggesting, as he does so, how difficult it is to secure a world composed of harmonies alone.

First Coda

Jocasta, James Baldwin, and Tragic Possibility

Thebes and the Plague

Sophocles lived from 497 to 406 BCE in Athens, his long life overlapping with the lives of Socrates, Plato, Thucydides, and Prodicus. *Oedipus Rex* was performed in 427 BCE, a few years after a devastating plague had hit Athens and a short time before an oligarchy enacted a coup. The plague killed close to 100, 000 people—25 percent of the city's population—and was accompanied by unrest. Some investigators today suspect it was typhus. The plague, a long war with Sparta, and a coup set a time of considerable duress and uncertainty.

Here is what Thucydides, a contemporary of Sophocles, says about that plague in Athens:

> Not many days after their arrival [of the Spartan invaders] in Attica the plague first began to show itself among the Athenians. . . . A pestilence of such extent and mortality was nowhere remembered. . . . Supplications in the temples, divinations, and so forth were found equally futile. . . .
>
> After the second invasion of the Peloponnesians a change came over the Athenians. Their land had now been twice laid waste; and war and pestilence at once pressed heavy upon them.[1]

Thucydides refuses to invoke gods as efficacious beings in that account, thinking that reference to them offers neither explanatory nor curative powers. His vision of tragic possibility reflects a sense of strange coincidences of timing in life, sometimes including untimely intersections between social and nonhuman trajectories, like a plague pouring from outside into human life or an eclipse that encourages a general to make a fateful delay in attacking an adversary.[2]

Sophocles added more characters to the tragic poetic tradition, with struggles between them about the role and beneficence of the gods integral to the rapid turns encountered. His own views, if they do emerge, do so through the dramatic interplay between events and disparate characters who contribute and respond to them.

Oedipus Rex, set in Thebes, unfurls a tragic result—a king who inadvertently killed his father, married his mother, sired four children in incest, and placed an entire regime at risk. Its diverse characters make different contributions to the result; they also offer contending interpretations of it. Was it due to a fate preordained by angry gods, the hubris of Oedipus in relation to the gods, a series of untimely events joined to Oedipal hubris coming together at the wrong time? The more characters, the more happenstances of timing enter into the mix.

Oedipus had saved the city from a plague fifteen years earlier when he—fleeing in panic from Corinth because of a distressing prophecy blurted to him by a drunken stranger—reached the gateway of Thebes. A Sphinx blocked his entry. He solved the riddle the Sphinx posed to him. The Sphinx plunged to its death in disappointment. Oedipus enters the town, soon marries the queen, and becomes king.

Thebes, following fifteen years of prosperity, now finds itself once again in the midst of a devastating plague. Can Oedipus, who rescued them once, save them again? Or is he part of the problem, as a few grumblers who recall the past may suspect ever so quietly?

A priest reports the plight of the City to Oedipus:

> The crops diseased, disease among the herds,
> The ineffectual womb rotting with its fruit.
> A fever-demon wastes the town
> and decimates with fire, stalking hated
> through the emptied house where Cadmus dwelled . . .
> We know you are no god . . . that is not why we throw ourselves before you here . . .
> It is because on life's unequal stage
> we see you as first of men and consummate
> atoner to the powers above . . . [3]

A double message. The suffering citizenry revere him as the "the first of men"; but, in these dire circumstances, he had better deliver the goods or the grumbling could grow louder.

The early speeches of the confident Oedipus fill the stage with double and triple meanings: meanings he intends; those that intimate the true object of the royal hunt; and those that jostle undercurrents of unrest and uncertainty in the subjects.

> Perhaps one of you is aware
> the murderer was someone from some other land.
> Let him not be shy to say it.
> I shall heap rewards on him,
> besides my deepest blessing.

And soon:

> That man, whoever that man be,
> I this country's reigning king
> Shall sever from all fellowship of speech and shelter,
> sacrifice and sacrament . . .

He chastises the city for not hunting down the killer of their previous king. That sin, he insinuates, could even be a source of their suffering now:

> Such ties swear me to his side
> as if he were my father.
> I shall not rest until I've tracked the hand
> that slew the son of Labdacus.[4]

"As if he were my father." The hunt is on. The king of Thebes knows himself to be the hunter. But do some elders sense vaguely that Oedipus even looks a bit too much like their former king, Laius? Things get worse when Tiresias enters the scene to accuse the king of being the object of his own hunt.

As the search and civic suffering unfurl together, Jocasta, the queen, seeks to reassure her handsome, young husband. She herself had once believed the gods intervene in human affairs; she had even sent her infant boy to be killed after the gods warned that he was destined to kill his father and marry his mother. The failure of that prophesy, she believes, reveals that there is no such divinely ordained chain of tragic causality. She now advances a different view of the world, rather close to one we have encountered in Prodicus. She seeks to reassure her now anxious husband—even as she notes he looks a bit like Laius as his hair has begun to gray along the temples—that Laius was killed "where three roads meet" by several brigands yet to be found. So no need to worry that he could have killed his own father, even inadvertently.

A crossroads? Oedipus is shattered by a sharp jolt in his gut. An affect-laden premonition has formed, powerful but not yet consolidated. For he, too, had killed a man at a crossroads, as he wandered from Corinth to Thebes to escape the drunken man's warning that he was doomed to kill his father. He now senses that he may be the target of his own search. "My queen, each word that strikes my ear has shattered peace, struck at my very soul."[5]

Each of us, too, has felt such a thud in the gut at dark moments, when someone says something arousing new doubts, inciting vague memories, or activating a suspicion previously simmering below conscious attention.

Sophocles knows us to be layered beings, with cloudy dimensions of memory and future projections simmering within more overt recollections, sometimes in tension with them. That is one reason those speeches with double and triple meanings cut so powerfully into us. They tap obscure traces of memory and suspicion in the audience below recollection. Oedipus himself is now poised in uncertain malaise. He now pursues the trail further to subdue a trauma that won't go away. This is the royal route pursued to save the city, and now, to resolve emerging doubts about his own regal sovereignty.

Jocasta reluctantly agrees to the search's continuation, but, while doing so, she also prods her young husband and king to consider a different reading of the cosmos they inhabit. To relieve the clamor in his soul. The cosmos, she says, is not governed by powerful, capricious gods who love to screw with the fate of human beings. That worry can be subdued if you overcome the view—the common Theban view she once held herself—that much of life is preordained by gods who often treat human beings as playthings:

How can man have scruples
    when it's only Chance that's king?
There's nothing certain, nothing's preordained.
We should live as carefree as we may.
Forget this silly thought of mother-marrying.
Why, many men in dreams have married mothers,
And he lives happiest who makes the least of it.[6]

This statement deserves unpacking. First, it shows Sophocles to be acutely aware that some Athenians did not accept common stories about the gods. Second, it shows that some of them assumed that if and when this weight is lifted from the world, a tragic vision would also be subdued. Jocasta thus welds together two themes: rejection of the power of the gods and relief from tragic possibility. Thucydides would, I imagine, disagree with her on the second count. He *connects* tragic possibility to a world replete with unexpected contingencies of multiple sorts.

What if a case can indeed be made to accept some version of the first Jocasta claim while reworking the second in the direction of Thucydides? The enhanced role of contingency she celebrates (after lifting the power of gods from life) does not mean humans are in control. It means, rather, that we are periodically susceptible to fateful concatenations. If contingent conjunctions play an enlarged role in this cosmic vision, humans may inhabit a social world and cosmos replete with both rich potential and tragic possibility. Contingently timed conjunctions between diverse events can sometimes create a blessing, sometimes difficulties, and sometimes catastrophes. As when, say, the conquering Oedipus and grieving Jocasta met in Thebes after he subdued the Sphinx. Or when a plague invades the city.

Does Jocasta herself, indeed, eventually adopt such a view, maintaining her dismissal of the gods but now embracing a vision of tragic possibility through strange conjunctions? We can't say for sure. But in an era of climate wreckage and viral species crossings in a world now tightly connected through air, train, ship, and car travel, fascist danger, and renewed risks of nuclear holocaust, it may be wise to treat that as one interpretation to mull over.

Jocasta may have rescinded her chancy/carefree view just before committing suicide in favor of a return to divine ordainment. Or, she may possibly have dropped the carefree tone and adopted a view of tragic possibility through a concatenation of bad human decisions and contingent nonhuman events. Killing herself because of the incredible shame that was coming her way. So, let's continue.

Oedipus summons the messenger. He, under regal pressure, confesses that he gave Oedipus to a shepherd years ago, disobeying the royal order to kill him. Now the murmuring from older people in Thebes becomes more understandable.

When Jocasta hears that the shepherd had been given an infant in a mountain dell of Cithaeron, it becomes her turn to turn pale. That mountain is the very one Dionysus, the god of wine and an element of chaos in the cosmos, inhabits. So, when she and Oedipus are informed that the infant's ankles had been riveted by a pin, Jocasta loses it. The name Oedipus means "Swollen Foot."

Yet another character to turn events in this drama: Oedipus, Jocasta, Tiresias, the priest, the chorus, the messenger, and a shepherd on the way. The multiple characters give the drama its complexity, its rapid turns, and its persistent aura of foreboding.

Time itself now seems to be out of joint, with its trajectory turning first this way and then that, as the pace of events accelerates.[7] Jocasta struggles to halt the spiral. She knows the jig is up if Oedipus interrogates the shepherd. "Why ask?" she implores the hapless man with the swollen foot. "Forget it all. It's not worth knowing."[8]

Not worth knowing? Has Jocasta silently shifted from being carefree in a world punctuated by chance events to acknowledging that a world laced with contingencies can itself sometimes manufacture grotesque results? Could, perhaps, the plague haunting the city drift away gradually on its own if *this* secret remains unveiled? Or has she returned to her earlier gods-centered view? We don't know. All we know for certain is that when Oedipus, the previous master of riddles, now insists on pursuing this riddle to its end, she laments: "Goodbye, my poor deluded one, lost and damned! There's nothing else that I can call you now."[9] She abandons her "deluded" son/husband.

Key secrets are also lost when Jocasta, the lovely, thoughtful matriarch married to her son, is lost to suicide.

The messenger tries to squeeze the truth out of the old shepherd, while Oedipus stares down both of them. As rulers do. The shepherd screams at the messenger, "Damn you, man! Can you not hold your tongue?"[10] The old shepherd, too, like Jocasta, may now think that the best chance to hold the regime together at this delicate point is to cover up the true identity of Oedipus. Does he too, perhaps, doubt that the gods will punish him for that refusal? But Oedipus squeezes the truth out of him. The shepherd admits he had been given the infant of Jocasta and Laius to kill, in their desperate attempt to outwit the gods.

To protect the baby—perhaps because he too doubted at some level the all-controlling power of the gods?—the shepherd made a courageous decision to pass on the crippled infant to another man, the ruler of Corinth. He secretly disobeyed his rulers to save the infant. Several characters in this drama take such surprising initiatives, outflanking a few readings of Greek tragic dramas that they deny characters "free will."

Now Oedipus encounters his fate:

Lost! Ah lost! At last it's blazing clear.
Light of my days, go dark. I want to gaze no more.
My birth all sprung revealed from those it never should,
Myself entwined with those I never could.
And I the killer of those I never would.[11]

It is all over. Oedipus is revealed to be the king who killed his father, married his mother, sired four children through incest, and placed an entire regime at risk. All his actions are now soaked with profound shame—a feeling of disrepute and unworthiness connected not to guilt over intended acts but to ugly reversals of his highest aims.

Rules and norms normally thought to fit together in a good Greek regime—monogamy, an incest taboo, rule by a wise king—now become disconnected, because of either the gods, the hubris of a ruler, or a strange set of untimely conjunctions. Oedipus now says, after blinding himself in grief and shame: "The gods? They are my enemies."[12] We also know what Creon roars—the brother of Jocasta who is now to assume the throne: "Stop [Oedipus] this striving to be master of all. The mastery you had in life has been your fall."[13] Oedipus concludes that his fate was preordained by hostile gods who treat individuals and Greek city states as "playthings." Creon affirms the beneficence of the gods, and puts the blame for these tragic events at the feet of Oedipus himself, the agent of hubris who once saved the city from plague and later doomed it to the same thing.

What, though, about Jocasta and the shepherd? Their distinctive voices were silenced before uttering final thoughts. Do they finally come to accept this tragedy as preordained? That is perhaps probable. But could there also have been a subterranean voice in each that links tragic possibility to a set of untimely concatenations between human and nonhuman events, forged in this case through bumpy intersections between a plague, Oedipal hubris, contending actors, a populace quick to blame bad outcomes on disobedience to the gods, and a turbulent, larger world not reliably predisposed to human welfare? Might Jocasta and the shepherd have moved closer to Thucydides than to Tiresias? That is the voice, at least, that I seek to *extract* from this play and carry to today. An image of tragic possibility as an *emergent* phenomenon finding periodic expression in fateful concatenations between planetary events and rash public decisions.

Jocasta was a Theban queen; James Baldwin an African American, gay man growing up in poverty who made a go of it in a racist culture. Each may give voice, across impressive differences of place and time, to a vision that speaks simultaneously to the time of climate wreckage.

Another Country

I grew up in the white working class in Flint, Michigan, during the 1950s. My parents, at odds with their own larger families on this score, were adamantly opposed to "prejudice"; but they did not grasp the deeper sources and institutional supports of racism. I am not sure I have sufficiently plumbed them myself.

I did see firsthand how the daily grind of factory work wore many white workers down, how some men carried these frustrations and disappointments home to abuse their wives and children, and, above all, how many—particularly those tied to white evangelical churches—clung to a fragile sense of identity by embracing fervently superiority over Blacks, immigrants, non-Christians, women, and Amerindians, for starters. It was not a surprise, then, when I encountered Georg Simmel's account of how a stratified system tends to be self-sustaining, with each stratum striving to pass to those below it the troubles and insecurities it receives from the stratum above it. Here is what Simmel says:

> This is the tragedy of whomever is on the lowest rung. He not only has to suffer the deprivations, efforts and discriminations which, taken together, characterize his position; in addition, every new pressure on any point whatever in the superordinate layers is, if technically possible at all, transmitted downward and stops only at him.[14]

Nonetheless, I was in for a shock when I read *Another Country* by James Baldwin in 1960. The novel, set in Harlem and the Village of the 1950s, explores crossings and impasses between whites and Blacks. It is electrically charged. The multiple pressures and ambivalences absorbed into its characters and relationships become almost impossible to bear. Rufus, a Black drummer in a jazz band, has fraught relations with white friends, particularly Vivaldo. When he meets Leona, a white woman who is herself a refugee from the south, he is simultaneously drawn to her, angry at her, and distressed at himself over that ambivalence. She, too, is loaded with ambivalence, having escaped an abusive husband in the south and lost custody of her child.

As they walk together down the street after another rocky night, he reviews how high the price of this racial crossing is: "trouble with the landlord, with the neighbors, with all the adolescents in the Village and all those who descended during the week ends. And his family would have a fit."[15] His sister Ida would insist that to continue this fraught relationship with a white woman is to hate himself and his race.

His abstract anger at white domination finds expression in ugliness toward Leona, in the way, for instance, he has sex with her. His penis becomes both an organ of desire and a weapon of racial conflict.[16] All this trouble in "the Village—the place of liberation."[17]

His drinking worsens and he loses a steady job. She collapses and is taken to a mental center; he is not allowed to see her. His white friend, Vivaldo, now becomes an adversary. He now treats Rufus as "a case."

Boarding the subway at night, Rufus observes, as if from a distance, "many white people and many black people, chained together in time and in space, and by history, and all of them in a hurry. In a hurry to get away from each other . . . but [they] ain't never going to make it."[18] Whites and Blacks, entangled in stratified crossings and ambivalent entanglements can readily descend into violence. Rufus stays on the subway beyond his stop, continues uptown while observing the class/race system in living color, until he reaches the stop uptown near the George Washington Bridge. He climbs the stairs and jumps into the darkness.

The closing scene of that chapter may pull you back to a few lines in an essay Baldwin had published a year earlier. There, he recalls a moment in 1943, when he was about to attend the funeral for the stepfather he had struggled with and against. He entered a restaurant in Princeton with a white friend. Once again he was told "we don't serve Negroes here." Something cracked.

> Somehow, with the repetition of that phrase . . . , I realized that she would never come any closer and that I would have to strike from a distance. There was nothing on the table but an ordinary water-mug half full of water, and I picked this up and hurled it with all my strength at her . . . and it missed her. . . . A round, potbellied man grabbed me by the nape of the neck just as I reached the doors and began to beat me about the face. I kicked him and got loose and ran into the streets. My friend whispered, "*Run!*" and I ran.[19]

Baldwin himself, it becomes clear, struggled repeatedly against the despair and hatred that had consumed Rufus in another country.

The story of another country lurking in this one proceeds for three hundred more pages. People complicate, surprise, and collide with one another across stratified systems of color, gender, class, faith, desperation, occupation, and age.

In a racist system, everybody becomes damaged goods. Webs of tragedy are woven out of such disparate threads between larger forces and personal stories. As Baldwin knows.

The Fire Next Time

When we encounter *The Fire Next Time*, published in 1962—a memoir of sorts woven into accounts of the American racial scene—it becomes apparent how Rufus lives on in Baldwin. But Baldwin also transcends him as he continues to think about the tragic.

Let's look first at the early life of James Baldwin, noting the binds he faced and the razor-lined labyrinth he negotiated to cope with them. His home life was stressed by an angry, punitive stepfather; the neighborhood was occupied by whores, pimps, and racketeers. When he brought a Jewish friend home one day, his dad hit him for doing so. He told his dad that his friend was a better Christian than he.

He attached his hopes to a Black evangelical church in the area. At the age of fourteen he became a preacher. He found a voice there, devoting it to warning parishioners of eternal damnation if they deviated from the narrow path of Christian salvation. But increasingly he became aware how treacherous such a path was: you cannot really sustain it under the duress of circumstances, and you were condemned if you do not. Even preachers were pulled into ugly deceptions. "I felt that I was committing a crime in . . . telling [people at church] to reconcile themselves to their misery on earth in order to gain the crown of eternal life."[20]

He left the ministry, looking for a new line of flight. He explored the Black Muslim movement and broke with it, too. Its nationalism seemed suspended in a void. Where would this nation be located? How would the dogma invested in it avoid becoming the inverse of the white dogmatism it opposed? In reviewing these moves, Baldwin always strives to grasp the pressure cooker that propels those looking for diverse escape routes. And to expose the traps in each flight path.

Baldwin was not endowed with the resonant voice of a singer, nor did he play a musical instrument, nor have magical athletic skills to parlay. The talents that provided hopeful escape routes for Blacks in that decade were not routes available to him. He was poetic, charismatic, and a talented writer. He now becomes James Baldwin, the brilliant author who insists upon speaking to both Blacks and whites. He strives, too, to perfect his voice as one who does not conform to white, heterosexual norms. A double whammy. If two young men walked down the street holding hands in my neighborhood during this period they would have faced the dangers Rufus and Leona had in theirs.

Baldwin insists that whites need to figure out what they are defending and what they lose through white triumphalism. But he does offer a few hints.

"White people guard too much," he says.[21] What do they guard? They guard their identities as people on the road to historical progress through capitalism; they guard their identities as white people; they guard their faith in life after death. They think they are guarding themselves, but they are guarding a vision that cannot be sustained much longer. Hence, they are apt to become tighter and more truculent as the elements that hold this identity together start to falter. "It is rare indeed that people give. Most people guard and keep; they suppose that it is they themselves and what they identify with themselves that they are guarding and keeping, whereas what they are actually guarding and keeping is their system of reality and what they assume themselves to be."[22] More generally, "white Americans have supposed 'Europe' and 'civilization' to be synonyms—which they are not."[23]

So, in guarding and keeping their identities, whites guard a system that cannot long continue to sustain itself. For example, the sharp divisions between lower-class Blacks and whites help capitalism to maintain low wages and high job insecurity. If differentially, for both.

Baldwin's Emergent Tragic Vision

It is here that Baldwin introduces a tragic vision of possibility. He drops the Greek gods from the cosmos but retains the idea that larger forces are both densely interwoven with human life and not inherently predisposed to their aims or welfare. That image is defined during a period when the Cold War was firmly in place. "Behind what we think of as the Russian menace lies what we do not wish to face, and what white Americans do not face when they regard a Negro: reality—the fact that life is tragic. Life is tragic simply because the earth turns and the sun inexorably rises and sets, and one day, for each of us, the sun will go down for the last, last time."[24]

That last statement deserves unpacking. To me it means, first, that to Baldwin there is no life after death. The task, to him, is to come to terms affirmatively with death as part of the wonder of life itself, without demanding eternal salvation for some and damnation for others. It means, second, that white unconsciously absorbed assumptions that the trajectory of the current civilization will continue into the indefinite future are not grounded in the nature of things. Things are much more fragile than that. A regime might encounter internal limits or trajectories from elsewhere that challenge and unsettle it. Or both together. It means, third, that you can't count on either a providential design or the receptivity of nature to capitalist mastery to continue indefinitely. That which appears to be solid is inhabited by more fragilities than commonly admitted or imagined. If only "because the earth turns and the sun inexorably rises and sets."

If and when such a tragic vision of possibility sinks in, it is tempting to become profoundly resigned or aggressively nihilistic. Indeed, the latter responses currently grip the cultural unconscious of many white evangelicals and neoliberals today. The seeds of white superiority and fascism are sown by such denials and demands.

But it is also possible, as more constituencies work to overcome what might be called white optimism and Christian providentialism, more will also come to appreciate the fecundity of life, along with the dangers, explorations, and experiments it foments. Baldwin links acceptance of a tragic vision of possibility to cultivation of more open and experimental approaches to ourselves and civilization. "One is responsible to life: It is the small beacon in that terrifying darkness from which we come and to which we shall return."[25] How does that work? Upon affirming tragic possibility as a condition of being, you work upon yourself and others to amplify love of life and to cultivate presumptive generosity to others. A philosophy of tragic possibility as an emergent phenomenon issues no guarantees. Its advocates, rather, strive to strike chords of resonance in others as they seek political alliances with those from different onto-cosmological orientations who also care for life.

Baldwin and Climate Wreckage

We can be confident that James Baldwin, were he alive today, would incorporate massive evidence of accelerating climate wreckage into that vision of tragic possibility. Why? Because it adds to the picture he has already drawn of fragilities and exposure to nonhuman processes haunting the trajectory of class and racialized capitalism. Just as "the earth turns and the sun inexorably rises and sets," *today conjunctions between a stratified capitalist system and a variety of nonhuman forces and agencies periodically render life for those on the bottom of the race and class systems even more fragile.*

Today, as extractive capitalism pours huge amounts of CO_2 and methane into the atmosphere, the emissions activate a series of planetary *amplifiers and distributors* that both render the consequences more severe and distribute them disproportionately to racialized nontemperate zones that have generated the lowest emissions. For instance, as climate heating increases the number, duration, and intensity of hurricanes in the Caribbean, populations in the Caribbean and American south are pelted by them, increasing the suffering of people on the islands, Mexico, and American south. Such examples of impersonal planetary carriers of imperial, racist, and class power are pursued in greater detail in the last chapter.

A tragic vision of possibility, to me, focuses on a fragility of things that, when acknowledged, undermines complementary cultural insistences that either providence or capitalist pursuits of mastery over the world carry reliable existential guarantees. By tragic possibility I mean a world that is neither highly susceptible to mastery as secularists often assume nor as providential as Christians often enough assume. Neither/nor. To come to terms with tragic possibility is not only to accept the fragility of things; it may also be wise to rework the very images of time in which many western secular and Christian images have been set. It is not that *all* the participants in such mobilizations must embrace a tragic vision of possibility. But it may be wise for more of us to do so. Baldwin, for example, embraces both care for life and tragic possibility. It is possible Jocasta did the same. Two emergent philosophies of tragic possibility.

I let Baldwin have the last word here, as he ponders the temper to cultivate in a world of tragic possibility:

> White Americans do not understand the depths out of which such an ironic tenacity comes, but they suspect that the force is sensual, and they are terrified of sensuality and do not any longer understand it. The word "sensual" is not intended to bring to mind quivering dusky maidens or priapic black studs. . . . To be sensual, I think, is to respect and rejoice in the force of life, of life itself, and to be *present* in all that one does.[26]

2

Augustine and the First Conquest of Pagans

Greek and Roman thinkers such as Hesiod, Plato, Socrates, Sophocles, Thucydides, Aristophanes, Epicurus, Prodicus, Ovid, and Lucretius, of course, did not call themselves "pagan." *Paganus*, the Latin word, was applied to rustic Romans before Christianity prevailed; it then became an epithet deployed by Christians in Rome to characterize a diverse cohort of non-Christian orientations to the world in Egypt, Greece, Rome, Persia, and elsewhere. Most of these "pagans" did not confess a universal, personal, omnipotent god who promises believers the possibility of eternal life in heaven if they obey key tokens of faith and receive its unfathomable grace, though some did confess an impersonal ur-god who started things off and was worthy of devotion. Most pertinent, Roman devotees did not confess that Jesus, born of a virgin, was himself resurrected after persecution, authoring a New Testament that rose above all earlier pagan and Jewish texts.

We begin with Augustine's *The City of God Against the Pagans*, completed in 413 CE; it was designed to defend and justify Christianity against pagans of all sorts. Rome had recently been sacked by the Visigoths. Augustine wrote a book of over one thousand pages, first, to defend against the charge that the recent conversion of Rome from paganism to Christianity was responsible for the decline of Rome; second, to explain why all versions of paganism were inferior to Christianity; third, to explain again (he had earlier, too) how the world was really created and what the universal god demands and expects of believers; fourth, to reveal what believers could hope for from this life and the next life; and, fifth, to advise how pagans should be treated in Rome. It was a massive undertaking.

I undertake this reading of *The City of God Against the Pagans* to identify definitive themes and conclusions in Augustine's text about pagans of his own day and, particularly, to show how his readings prepared the stage for a second great encounter with pagans when Europe invaded the "New World" after 1492 CE.

A couple of clarifications may be helpful. I do not write these pages either as a loyal Augustinian or as one who once was one and then "left the faith." I write as one who has been a nontheist for a long time and has felt compelled from time to time to refine a crude version—the atheistic version—that orientation too often takes. A nontheist seeks room in public space for their theo-philosophical stance, but does not demand that it monopolize this space. As a nontheist, however, I do risk missing some themes that those inside the faith capture best, as Augustine surely did with his take on pagans. I write, additionally, as an obdurate pluralist who seeks a world in which a plurality of final faiths intersect and interact in both the public realm and private life, each drawing upon its fundamental faith when pertinent to do so, each acknowledging without existential resentment that its own creed is reasonably contestable to others, and each seeking to contribute to a positive ethos of engagement across creeds through which general policies can be negotiated. Such a deep pluralism, of course, itself must draw boundaries. But its boundaries are different from those pursued by Augustine and many other uni-theists or uni-atheists after him. If and when, for instance, plurality is threatened by a singular movement toward white nationhood, Christian hegemony, or fascist unity, such nontheists seek to mobilize a pluralist assemblage strong enough to ward off the effort.

The pursuit of pluralism, of course, is replete with uncertainties and fragilities. It is just that, in my view, it is less violent, less oppressive, and once established, richer and less fragile than other extant alternatives in a world that always and already teems with protean diversities. More about those issues later. For now, it might suffice to say that I approach Augustinian texts as an outsider seeking to understand them in relation to pagan onto-theological orientations from which they diverged. Seeking to do so because of positive insights the texts offer, because of severe injustices they promoted in their own day and horrific contributions they made to future European invasions of pagans in the Americas. To mark my ambiguous relation to the text as we proceed I will place the word "god" in lower case (except when quoting Augustine and others), doing the same for pagan gods. This stylistic choice expresses a desire to place diverse onto-theologies on something closer to horizontal planes of discourse rather than along a vertical hierarchy of onto-authority.

A further word about the last concern noted above. I will not suggest either that Augustinianism *caused* the second conquest of pagans in the "New World" or that its themes were *irrelevant* to conquests undertaken for other reasons such as acquiring gold, strategic competition, new sources of labor and people to enslave, safety valves for domestic violent urges, new sources of capital, or earthly grandeur. Rather, I will suggest that nested inside those latter pursuits were (and are) a set of culturally engrained *prompts*, existential *demands*, and ontological *precursors* authorized by Augustine to secure the singular majesty of the Christian god. There is, of course, no culture without a set of embodied cultural prompts and bearings, as Augustine himself understood so well. But this drive to universalization of a particular set needs to be overcome by valiant intellectual efforts and cultural struggles. Indeed, you might say that a *third drive to* conquest—the attacks on Euro-American, pluralizing states by neofascist forces—expresses drives to revenge within settler societies against recent attempts to surmount old Christian/capital logics of internal closure and imperial capture.

What about these pagans, then? Except for a few Greek thinkers who were said to anticipate vaguely Christianity—Plato being the most important of them—Augustine found pagans to inhabit an imaginary world that was too disorganized, tumultuous, populated by uncertainties, and filled with diverse gods animating various aspects of nature—the sea, volcanoes, human passions, wine, the soil, love, lightning, and so on. That last theme of multi-vitalism insinuated into nature itself, perhaps, was defined to be the biggest sin of pagans, since it refused to reserve all creativity to a single god who created the world from nothing. Pagans were rural, provincial, and unattuned to a singular god who governs everything, including time itself.

Those pagan gods, with overlapping functional and territorial assignments, acted in confusing ways. Their heavenly world was too jagged and tempestuous compared to the smooth heaven Augustine confessed, one with a god whose omnipotence and omniscience qualified it alone to engender eternal life after the earthly death for those who receive imponderable grace. Pagans, to him, were unlikely to receive this award of providence from the single god. The gods they professed were too weak and divided to deliver it. That is a clinching point for Augustine, as we shall see.

So, to Augustine, pagan gods were "foul spirits," demons whom he surmised actually existed and could wreak havoc but lacked worthiness to be worshipped or even placated. Pagan rites were thus foul and filthy too.

> And these rites involved *fercula*, as though some feast were being celebrated at which the impure demons were to be delighted by a banquet given in their honour. But who does not see what manner of spirits they are who take delight in such filthy things?"[1] "This mother of the gods [Juno], then, was of such a character that even the worst of

> men would have been ashamed to have her as his mother. Yet, in order to take possession of the minds of the Romans, she sought out the best of men—not to admonish and help him, but to cheat and deceive.[2]

Immediately we see that Augustine appraises these gods according to a truth they lack. He does not appreciate how many followers sought to *placate* them because the gods themselves could unleash forces hostile to human beings, or that some even treated the gods to personify nonhuman forces that could bring good tidings to people at some times and disaster at others. The idea that there might be nuggets of wisdom in pagan appreciation of a world not intrinsically predisposed to humans themselves was existentially unacceptable to Augustine. A tragic vision of possibility defiles his god, an omnipotent being who, nonetheless, bears no responsibility for evil in the world. No, the Greek and Roman gods are judged by the extent to which they have the power and interest to bring eternal life to people. Lacking omnipotence, their power in this respect is deemed unreliable.

Augustine often repeats this charge against pagans: the very plurality of limited gods they projected could not ensure the possibility of eternal life. And eternal life is the golden seal to pursue. It was thus not merely that Constantine—the emperor who moved from paganism to honor Christianity and support the primacy of the Trinity—enjoyed more felicity in *this* life after doing so; "for every man should be a Christian only for the sake of eternal life."[3] The highest court through which to judge a religion, then, is the degree to which its ordinances enable the possibility of eternal life for those who embrace it. Pagan faiths fail on this count. So do the creeds of Manicheans, Arians, Pelagians, and Jews, for different reasons in each case. In some cases the creed does not pursue the singular aim of religion, in others the creed pursues it but lacks the theological machinery to deliver the goods.

To Augustine, attainment of any high degree of complexity—say, the capacity of humans to choose freely, to worship a god, to practice virtue—*can only find its source from a yet higher mode of complexity and power.* That is what I call the doctrine of *eminent causality*.

Some pagans, Epicurus, Lucretius, Thucydides, and Prodicus among them, held that high modes of complex life can *emerge* from the fecundity of lower processes. That is because lower processes bristle with loose potentials that exceed what they are and because they periodically make chancy contact with other such forces. Out of such conjunctions, higher forms might emerge. Modern theories of evolution are founded on that assumption, including the confidence that life itself emerged from modes of nonlife bristling with potentialities that exceeded what they were.

Eminence vs. *emergence*. Theologies of emergence usually start with the assumption that the flow of time and some earthly materialities preceded the emergence of higher complexities, including humans. Augustine found variants of this latter view to be outrageous, dangerous, sacrilegious, and insulting to the omnipotent god he confesses to be the singular, eminent source of all being, including human being, the human will, and time itself. Why, though, rather than marking off an important difference of faith—as Julian, the short-term pagan ruler of Rome had commended before Theodosius insisted that Rome be a Christian empire—was Augustine himself deeply committed to treat such differences of fundamental faith to be so insulting to his god and so urgently in need of authoritative curtailment?

The issue is not whether the earlier pagan persecution of Christians was to be condemned. It should have been and should be. But whether these faiths were to be defined so contemptuously once Christianity gained power. It is also not whether it was viable to hold the Augustinian faith in eminent causality as a noble, contestable view. It was and is. The issue is whether it was cruel and unnecessary to defile other existential faiths that denied such claims.

The first theme of Augustine's disparaging readings, again, is that the pagan faiths lacked the ability to appreciate a god able and disposed to promote eternal life. There is a second failing, too, but it is not stated so directly. Several pagan orientations treated the nonhuman world itself as not reliably and regularly disposed to human well-being. It releases events that create human suffering and turmoil, either because the gods provoke them, or because the gods are limited, or because the gods are taken to be figures for forces that can be volatile. Augustine quotes Apuleius from *De mundo*, a text no longer available to us. To Apuleius:

> all earthly things are subject to change, overthrow, and destruction. For indeed, to use his own words, "by violent tremors of the earth the ground has opened and swallowed cities and their peoples; whole regions have been washed away by sudden deluges; those also which had formerly been continents have been made into islands by the coming of strange floods. . . . Cities have been overthrown by wind and storm; fires have erupted from the clouds. . . . So also, rivers of fire kindled by the gods once flowed from the craters on Etna's summit and poured down the slopes like a torrent."[4]

These events, says Augustine, occurred "before the name of Christ had suppressed those rites of the Romans: those rites which are so vain and inimical to salvation."[5] The providential world was replete with terrible events before the birth of Christ, including the punishments of Adam and Eve and the later drowning of all humans who flouted the goodness of Noah. But as providential history hits its stride, such events become scarcer; if and when they do occur they occur as divine punishment for human sins. Thus, when an Antique Little Ice Age accelerated the tailspin of Rome 130 years after Augustine wrote his attack on the pagans, he would have treated the event as punishment for human sins rather than as either part of a rocky world or a world in which his god does not bestow its highest status upon human beings. Some readings of the Book of Job, indeed, do offer the second type of interpretation.

Thus two key Augustinian objections to paganism: its inability or unreadiness to define a divinity that can promise eternal life; its readiness to fill the nonhuman world with rocky events that testify either to a nonprovidential world or to godly limitation.

Let's review now the *tone* of Augustine's engagement with Marcus Varro, the Roman philosopher and humorist who revered messages from the Roman gods. He may well have treated the gods as figures to be admired and honored rather than beings to be taken literally. Though that judgment is uncertain, given the popular demand for belief that circulated through Rome when he wrote and the treatment writers who openly denied such beliefs could risk. He was born in 116 BCE and lived to be ninety. He wrote hundreds of books—most of which have been "lost." He studied in Athens for a while. We know that he warned people to avoid swamps populated by invisible creatures that could infect the body and cause disease. Emergent causality.

Augustine, unlike us, had access to Varro's work. He considered him to be a writer who had "outstanding" talent. Here are some other things he says:

> Yet he worshipped those same gods. Indeed, he esteemed their worship so highly that . . . he says that he was afraid lest they should perish not by the assault of an enemy, but by the neglect of the citizens. . . . Yet he presents to the world things to be read which wise and foolish men alike might rightly condemn as completely inimical to the truth of religion.[6]

> Varro himself attests that he wrote of human things first, and only then of things divine, because cities came into existence first and divine things were instituted by them subsequently. The true religion, however, was not instituted by any earthly city.[7]

> O Marcus Varro, you are without doubt the most acute and learned of men. But . . . you have not been raised up into truth and freedom by the Spirit of God.[8]

> Is anyone really content to ask or hope for eternal life from the gods of poets and theatres, games and plays? God forbid! Rather, may the

true God turn aside from us such wild and sacrilegious madness! . . . So, then: neither by mythical nor by civil theology does anyone attain eternal life. The former sows by devising wicked stories about the gods; the latter reaps by approving them. . . . The one sings of the crimes and disgraceful acts of the divine beings; the other loves them. . . . Both kinds of theology are vile, and both are damnable. . . . Are we to hope for eternal life from that which pollutes even this brief and temporal life?[9]

"Completely inimical," "such wild and sacrilegious madness," "wicked stories," no "hope for eternal life," "crimes and disgraceful acts," "vile," "damnable," "that which pollutes." This first line of Augustinian existential associations is contrasted to a second: "truth and freedom," "the truth of religion," "the true God," "eternal life." For several hundred pages Augustine repeats these themes about pagan authors, with numerous variations. The thought of Cicero, Plato, and Plotinus, for instance, though pagan, is ranked much higher than that of Varro and Epicurus. The latter denied either sufficient power or existence to the gods. The former were on the road to the true faith but did not reach it because they did not experience the life of Christ. Plato, for instance, found it probable that raw materials were already there when the earth and heavens were consolidated. Hence his *demiurge* is not omnipotent and cannot so reliably provide possibility of eternal life. Plato also lacked access to the inspiring, definitive, and personal word of Jesus Christ, the lord and divine being who will rise again to issue a Final Judgment on all human beings.

Augustine refused to ask himself a key question. Was the end of religion he thought to be the only true end worthy of universal pursuit—eternal life—itself so urgent an end that he had to scorn faiths that either did not honor it or did so in ways that did not invoke Jesus Christ? While some pagans—including Augustine's reading of Varro, and Julian (who was emperor for a brief period)—counseled a public plurality in which people inhabiting the same regime could honor diverse faiths with dignity, Augustine did not. He opposed the execution of pagans, but he did support the edicts that defined Christianity as the authoritative religion of the Roman Empire. Pagans could live in the territory, but their places of worship were often trampled with official sanction and their grievances were regularly demeaned.

But why do I, differing from Augustine in this regard, insist upon poring once again over the language through which he condemns and demeans paganism? This is the reason. The confession of Augustinianism as the official faith of an entire empire instills into future embodied prompts and precursors of European cultural judgment degrading dispositions against all lived creeds defined as pagan. For example, he does not ask whether it might be illuminating to learn more about pagan deities who personify the rocky, turbulent features of nonhuman nature, doing so because it is always wise to keep such turbulent possibilities in mind. Those gods, say, who personify raging oceans, thunder, plagues, lightning, storms, wild human passions, and so on, gods taken to be either *beings* or *figures* who teach about human relations to a rocky, nonhuman world. Moreover, he does not acknowledge that the faith with which he is imbued, and which he preaches as the Bishop of Hippo, is reasonably *contestable* to many others in ways he was unable to refute by impeccable means.

Most importantly, these very features of Augustinianism, sprinkled through a text of over one thousand pages, become inserted into the cultural unconscious of Christianized Europe through a host of disciplines and rituals to be examined soon, setting a stage for future invasions and holocausts that exceed them. His commitments to a providential nature and time and his treatment of pagans are two key reasons critical attention to his work remains so important.

Augustine found it to be a failing that sometimes in Greece and Rome diverse faiths and philosophies were tolerated in the public square. Such a "Babylonian" plurality constitutes an assault on the true faith. "Some asserted that there is only one world, others that there are innumerable worlds; some that this one world came into being, others that it had no beginning; . . . some that it is directed by a divine mind, others by fortune and chance. Some maintained that souls are immortal, others that they are mortal."[10]

To summarize the findings of this section we turn to the preface to book 7, where Augustine summarizes the judgments he has propounded to that point:

> I am here endeavouring most diligently to uproot and extirpate depraved and ancient opinions which the long-continued error of the human race has implanted deeply and tenaciously in the dark places of the soul; for these opinions are hostile to the truth of godliness. In performing this task, my own small ability is aided by the co-operation of the grace of the true God. . . . For we are here proclaiming a matter of the very first importance: namely, that the true and truly holy Divinity, even though He furnishes us with the help necessary for the frail life that we live now, should nonetheless be sought and worshipped not for the transitory vapour of this mortal life, but for the sake of the blessed life to come, which is nothing less than eternal.[11]

Before turning to positive articles of faith by which Augustine strives to help lift us to the true religion, let's look at disciplinary strategies he favored in his day to deal with wayward people such as rebelling sisters in a convent, resistant pagans, Jews, and Christian heretics who misrepresented the true character of the true faith.[12] As the bishop of Hippo, he had to make authoritative decisions on many issues, including regular disciplines to enact, and he campaigned to influence the entire church to apply them widely.

Disciplines Applied to the Faithful, Jews, Heretics, and Pagans

To gain a granular sense of the hierarchy Augustine embraces and enacts, we begin with his admonitions to a group of sisters who had complained to him about arbitrary rule by the mother superior in their religious order. He does not sympathize, as they had hoped, with their complaints. Rather, expressing concern that the voice of the devil may have captured their hearts, he counsels them to obey the mother superior and not to complain. "May God then calm and compose your hearts! May the work of the devil not gain the upper hand within you, but may 'the peace of Christ rule in your hearts'!"[13]

Clearly, the authority structure of the order means that the bishop must deploy complaints by sisters to internalize the hierarchy of authority: bishop, priests, sisters, everyday parishioners, those outside the faith. What the sisters identified as misuses of authority must be reinterpreted to be interior defects in the souls of the sisters in need of self-work. That is the Augustinian formula.

Augustine thus concludes that he needs to issue the sisters a more refined set of rules and disciplines to guide their daily conduct—authoritative rules to obey, handed down by the bishop. Each specification places the sisters in a potential bind: their conduct can now be judged more stringently by the mother superior, but it is also now more difficult to be confident about whether they are obeying the edicts. Are your habits of eye contact in public, for instance, too bold or too withdrawn? These binds are doubled when the sisters are then counseled not to question the mother superior, but to turn inward toward admission of their own wayward conduct when she admonishes them. Thus, a sister must not draw attention to herself by wearing her hair either too tightly or too loosely. The very uncertainty of the proper arrangement, clearly, now requires the sister to attend to herself more carefully; thus she now risks falling into another sin of too much self-attention.

The rules create new opportunities for the mother superior to chastise or counsel a sister to reexamine her own hidden, inner desires. Has she given too much attention to her hair? Too little? Are her eyes too forward? Too modest? Interiorized, subordinate, guilty selves are being composed here, through the hierarchy of authority, detailed rules imposed upon subordinates, and the questions they are commended to pose to themselves. Is the experience of a will divided against itself—a feature that Augustine construes to be part of the human condition itself after the fall—in fact being consolidated here?

The instructions by the bishop to the sisters continue. They include inconspicuous dress, walking in procession when in public to ensure that each observes whether the others are sufficiently modest, meditating with your heart while you pray aloud, and so on. Here is one statement:

> In walking, in standing, in deportment, in all your movements, let nothing be done that might attract the desire of anyone. . . . When you are in procession you are not forbidden to look upon men but to desire to make approaches to them or have them make approaches to you. . . . And do not say that you have chaste minds if you have unchaste eyes, because the unchaste eye is the messenger of an unchaste heart.[14]

Augustine is even harder on pagans and Jews who complain to him. At the time he writes, the church has become the official church of the empire. Augustine honors Emperor Theodosius for making it so. Jews are to be tolerated, but only as a despised minority whose early recognition of a single god was vitiated by their later, as he says, killing of Jesus and denial of the Trinity. They could never embrace Jesus. "But the Jews who slew Him and would not believe in Him, who could not believe that it behooved Him to die and rise again, suffered a more unhappy destruction at the hands of the Romans and were utterly rooted out from their kingdom. . . . They were scattered throughout the whole world."[15]

Here, as elsewhere, the passive voice does a lot of work for Augustine. It minimizes the agency of those humans who helped to uproot the Jews; it invests ultimate agency in a divine being whose actions either are now seen to be just or will be later after divine history has unfolded. For time itself, as it unfolds progressively, is invested with divine purpose and wisdom. Jews are on the wrong side of time.

Pagans faced sharp limits and disciplines too. Those residing in the empire are allowed to remain, but they are not well protected from official discrimination or popular uprisings. Augustine, as bishop, receives a letter from a group of pagans who complain about local harassment by Christians. His reply amounts to a warning that things will get worse unless they convert. He uses the current plight of Jews, the faith from which Christianity itself emerged, to provide worldly signs of Christian superiority, wielding their fate as a warning to pagans.

> You plainly see the Jewish people torn from their abode and dispersed and scattered throughout almost the whole world. . . . Everything has

> happened just as it was foretold. . . . You plainly see some of the temples of [pagan] idols fallen into ruin and not restored, some cast down. Some closed, some converted to other uses . . . ; and you see how the powers of this world, who at one time for the sake of their idols persecuted the Christian people, are vanquished and subdued by Christians who did not take arms but laid down their lives . . . ; and you see the most eminent dignitary of this noble Empire lay aside his crown and bow in supplication before the tomb of the fisherman Peter.[16]

Here, as in other instances noted, Augustine invokes a passive voice to characterize a historical fate befalling those who do not recognize the true faith. Sure, there were often human vehicles of these violences, dispersals, scatterings, vanquishings, and subduings, but they were carriers of a divine will that exceeds them. It is divinely inspired time working itself out. We will encounter the repetition of such a passive voice of historical destiny in later Christian and secular encounters with pagans on other continents.

Lay Catholic women, living outside the religious orders, were also construed to be tainted with blame for misfortunes they suffered—including beatings by their husbands or rape by invading soldiers. Yes, many subjected to force were indeed innocent.

> But let not such women, even if the lust of the barbarians was forced upon some of them, bewail the fact that this was permitted [by a graceful god]. Let them not believe that God is neglectful of them because, in their case, He has permitted a sin. . . . For it may be that some of the heaviest loads of guilty lust are overlooked by God at present, yet reserved for a last and visible judgment. Moreover, it may be that these women, even though of good conscience because they did not allow their hearts to be swollen with pride in the virtue of chastity, nonetheless possessed some latent infirmity which might have grown up into the arrogance of pride had they escaped this humiliation during the sack.[17]

"It may be . . . ," "nonetheless possessed some latent infirmity . . ." In the Augustinian world of confessional selves who seek to decipher hidden tendencies to sin in their souls, men, women, pagans, priests, and heretics are all susceptible. But the weight of suspicion regularly falls most heavily on those lower in the divine hierarchy of being. Both the purity of the god and the hierarchy over which it presides are protected and legitimated through such tortured judgments and disciplines.

Augustine reads the historic dispersal of Jews and defeat of pagans—both events that required oppressive imperial power to succeed—to be a feature of history guided by the hand of god as the world advances toward a future of Christian achievement and Final Judgment. It could be worse. He could have threatened the complainers with violence or exile. And it could have been better. He could have pursued a course in which such creeds and faiths are welcome alongside others, as Julian, the pagan emperor, had done briefly—before his suspicious death. He had welcomed "Galileans" as one faith among several others in the empire.

Again, the dispersal of Jews becomes a salutary mode of warning to them and others. "They were scattered throughout the whole world (for there is certainly nowhere in the world where they are not present); and so, by their own Scriptures, they bear witness for us that we have not invented the prophecies concerning Christ. . . . Indeed, a prophecy concerning this scattering was given long ago in the Book of Psalms."[18] Many people must be left by the wayside as providential time works its convoluted course through history.

What about those within the church who dissent strongly from Augustinianism? Augustine identifies several heresies, abominable readings of scripture that both deserve to be excluded and, through their visibility, help to sharpen the very definition of the true creed. For bad things in the world have some underlying, beneficent effect over the long run of singular time, even though we do not always know what it is.

Augustine thus embraces Catholic definitions of Arians, Donatists, Manicheans, and Pelagians as heretics, as proponents of doctrines to be excluded from the church. He had himself been a Manichean for nine years, while he lived with a young concubine. But he was converted. The Manicheans resolve the problem of evil by identifying two opposed forces in the world. But this very divine duality means there is not one single god powerful enough to ensure the possibility of eternal life, let alone create the world from nothing. It must be defined as a heresy to be purged from the church.

Let us here attend to his recommended treatment of Pelagians, those who believed, with the priest Pelagius, that devotees can earn their way to heaven by free will and do not necessarily require the intervention of divine grace to enter that realm. Such a doctrine was very popular at the time. Augustine insists, late in life, that its proponents deny belief in original sin, as they did. That doctrine thereby diminishes the indispensability of grace, and it thereby calls the omnipotence of god into question. For must not entry into heaven be due entirely to god's decision and not reducible to human efforts that implicitly weaken his majesty? Such an effect, Augustine contends, will eventually call the very centrality of god into question.

He does not merely *debate* this doctrine under the canopy of the official church. He demands that those who have supported it be banished as heretics unless and until they confess humbly in public how wrong they have been. Here is what he says when pleading with the bishop of Rome to deal severely with those who have now lapsed into silence:

> Some . . . have suddenly become silent, so that it is impossible to ascertain whether they have been cured of it unless they not only refrain from uttering those false doctrines, but actually take up the defense of the contrary doctrines with the same fervor they showed in propounding error. These, however, surely call for milder treatment: what need is there to terrify them when their very silence shows that they are terrified enough? At the same time they are not to be passed over and spared remedial attention . . . because their sore is hidden. For . . . yet they ought to be taught, and, in my opinion, this process is easier while the fear they have of severe measures assists him who teaches them the truth.[19]

"While the fear they have of severe measures." Later, carriers of the Inquisition will show mild heretics the instruments of torture without actually using them. It is not that Augustine is responsible for such later practices and the extremes of torture that were invoked. Rather, it is that the definitions of heresy and practices of forceful induction or marginalization he embraced in his day take a step toward institutionalization of more extreme practices later. The demand to confess what the church insists you believe, so that the belief becomes more absorbed into the visceral prompts imbued in you, sets dangerous precedents of control. In much more radical ways, such a doctrine informed the horrific Stalinist show trials of a later age.

The profound Augustine knows that widespread, public confession of a doctrine is a powerful way to instill it more deeply into the wider culture and to pour it into the souls of practitioners. Some will thus say that his orientations to the sisters, to Jews, to pagans, and to heretics of different sorts were tame by comparison to earlier Roman practices against Christians, or that they were needed by any creed that seeks to define itself in public life. I accept the first point and disagree with the second. This was a moment when the church could have both welcomed more internal diversity within its faith and displayed much more respect for pagans and Jews outside it. Even more fundamentally, I worry about what such practices, once absorbed into the official life of the church, meant for later encounters with pagans on other continents not anticipated during the era of Augustine.

Is not the most pertinent thing, though, to report how the bishop of Hippo promulgated a positive doctrine and inspired others to absorb it? Yes. We turn to that now. Without forgetting that the presumptive definitions and treatments you bestow on those outside or on the lower reaches of the existential hierarchy you embrace secrete a pattern of authorization through which members of the faithful treat them, as well as how you, as a creedal authority, respond to entreaties when the treatment is severe. The question: Do the creedal inspirations, evidence, and confessions adduced in favor of this political theology vindicate the disciplinary methods, hierarchies, exclusions, disciplines, and modes of surveillance that consolidate it? That is, perhaps, a pagan question.

The Confessional God

Augustine's god is grounded on a mix of arguments, confessions, and institutional enforcements. The confessions acquire new energy each time self-doubt arises in Augustine himself about a feature of the doctrine. Now a new level of confession rolls forth, folding the terms of faith more deeply into the Augustinian self. The reader is called upon to follow along.

The *Confessions* were written after Augustine had flirted with Epicureanism, a doctrine that advised people to become more ethical by affirming mortality as a condition of life itself, and after he had pursued a nine-year affair with Manicheanism. Epicureans and the Manicheans, he later concluded, had one thing in common across their fundamental differences. Neither was able to found a creed that promised the real possibility of eternal life—salvation. That goal, Augustine insisted, is the true aim of religion itself. The first doctrine faltered because it construed earthly death to be final. To overcome resentment of this obdurate condition of life itself is to ease torment in your soul while living; it thereby helps you to infuse more care for others into life. The second doctrine failed because its division of the world into two contending spiritual forces—the forces of good and evil—meant that it was unlikely that a good, providential god would have enough *cosmic power* to deliver eternal life to the deserving. Neither could resolve the problem of evil to Augustine's satisfaction. The Manichean god was too limited to provide this assurance; the Epicurean faith made too many concessions to suffering.

Let's pick up the *Confessions* after Augustine confesses how he overcame those two creeds and still strives with all of his might to resolve the problem of evil. We listen as he confesses to his lord his creed and the zone of ignorance that still clings to it:

> By humble devotion return is made to you, and you cleanse us from our evil ways, and are merciful to the sins of those who confess to you, and graciously hear the groans of those shackled by sin, and you free them from chains that we have made for ourselves.[20]

Confessing to his god in a devout, obedient way, as he does throughout that text, Augustine seeks to strain humility and responsiveness to divine grace more deeply into his soul. But mysteries and uncertainties continue to haunt him. The biggest is how evil could persist in a world that this benevolent, omnipotent, omniscient, salvational god had created from scratch. Perhaps we should listen, then, to a key confessional moment in the past, when continued uncertainty about the source of evil meets his new certainty about the essential characteristics of god. He speaks to his god while confessing the essential characteristics he knows it to have.

> But now, O my helper, you had freed me from my chains, and still I asked, "Whence is evil?" but there was no way out. Yet in none of these wavering thoughts did you let me be carried away from that faith in which I believed both that you exist, and that your substance is unchangeable, and that you have care over men and pass judgment on them, and that in Christ, your Son, our Lord, and in Holy Scriptures, which the authority of your Catholic Church approves, you have placed the way of man's salvation unto that life which is to be after this death. These truths being made safe and fixed immovably in my mind, I asked uncertainly "Whence is evil?"[21]

This confession deserves close attention. First, it places at least eight articles of immoveable faith above review; it does so by confessing them devoutly into the soul. They are: god's existence, his unchangeable substance, his care over men, his authoritative judgment of them, his introduction of Christ into the world, the authority of Holy Scripture, the institutional authority of the Catholic Church, and the way of man's eternal salvation. To the extent these articles are grounded in confession, from whence derives Augustine's insistence that they must be treated by the empire as so superior to all other articles of faith that the Augustinian church must be given singular priority?

Second, it praises this god by expressing confidence that it has secured these elements of faith in Augustine through its divine grace, giving grace priority over choice, will, or decision.

Third, it closes by confessing the goal of "eternal salvation" to be the highest aim of the universal creed.

Fourth, the brilliance and existential dogmatism of Augustine coalesce in this passage: it is almost as if the brilliance, devotion, and sincerity of that confession suffices to obligate others to it. This indeed is not the last time such a combination comes together in western thought and institutional life, as we will see when we address Columbus, Sepúlveda (the harsh Catholic priest in South America), Descartes, and Kant.

Finally, a remnant of perplexity by Augustine remains vividly on display amid that confession of certainties: how to resolve the problem of evil within a faith that is now fixed in its key parameters.

After solidifying the parameters of his creed through confession—that is, by infusing them more deeply into our souls through devout repetition—Augustine tackles the problem of evil again, a problem that haunts not only Christianity but other creedal faiths. He proceeds along two tracks. The first is to assert faith that what appears to be evil during one period will turn out, when the Second Coming arrives, to have been part of a progressive historical pattern that weaves the greatest possible good into the world. The worry haunting that track is whether such a temporal unfolding in fact reveals this god to be limited in the early going. Later Christian nominalists will level precisely that charge against this historicist account, saying that faith in an omnipotent god means that you cannot ascertain the broad trajectory history and time will take.

The second track draws our attention now. It is the idea that the highest good of free will can only be sustained *if* humans—he means men above all—are free to will evil as well as, with the aid of god's grace, to will good.

Of course, most Greeks believed that intentional human agency is real. Even the elder Oedipus carefully explained to those still blaming him for his earlier acts when he approached the sacred grove in Athens, that he was an innocent, intentional agent. He had not intended to marry his mother, kill his father, have children through incest, or impel his wife/sister to suicide. His intentions had been otherwise and his actions followed from those intentions. But the gods intervened between intention and result to twist the results into grotesque shapes. For example, he intended to push that anonymous man out of the way at the fatal crossroad and then did kill him in a rage. It was the gods, however, Oedipus insists, who made the anonymous man be his father.

Augustine, however, makes free will depend upon the creative power only of his god; it only turns after the fall in the wrong direction. He also insists that divine grace is extremely important to what happens to people, whether, for instance, they go to heaven or burn eternally in hell. While he plays up the importance of free will in comparing his doctrine to that of pagans, his own version of fate is the doctrine of divine grace.

How could free will both emerge as the highest gift of divine eminence and participate in the lowest result—the production of evil? Augustine poses the question:

> Whence comes this monstrous state? Why should it be? Mind commands body, and it obeys forthwith. Mind gives orders to itself, and it

> is resisted. Mind gives orders to the hand to move, and so easy is it that command can scarcely be distinguished from execution. Yet mind is mind, while hand is body. Mind commands mind to will . . . but it does not do so. Whence comes this monstrous state? Why should it be?[22]

We note at the start that Augustine projects a version of mind/body dualism in which, first, mind or soul is imbued in us as the directing organ by god, and, second, the presence of soul sets a vast difference in standing and capacity between human beings and all other beings on the face of the earth. Animals are lowly and simple. Human exceptionalism, anchored in divine providence toward humans as its highest, unique source, is confessed here. Eminent causality allows god to infuse the will. Human exceptionalism, dualism, and eminent causality now run together in the same creed.

The stage is now set to explain how the highest human faculty—free will—can also be the site and source of the most evil performance. How could you will yourself to obey god but often sin anyway? The answer is that the first will to obey is polluted by a second will within the self, contending against it.

> But the complete will does not give the command, and therefore what it commands is not in being. Therefore, it is no monstrous thing [i.e., pagan thing] partly to will a thing and partly not to will it, but it is a sickness of the mind. . . . Therefore there are two wills, since one of them is not complete, and what is lacking in one of them is present in the other.[23]

Augustine seeks to subdue the problem of evil by relocating it in a division of the will within human beings who have inherited the original sin of Adam. Evil is a purely human affair, not grounded, say, sometimes in discordances between the priorities of a human culture and the periodic volatilities of nonhuman nature. Never grounded in the limitations or maliciousness of god either. For nature is the divine product of an omnipotent god, and god is providential. Augustine's god is thus saved from implication in evil, either through a lack of care for humans (which would threaten divine benevolence) or through limitation in its capacity (which would pose a threat to omnipotence). Now the single creator of the world cannot be held responsible for evils that emerge in the world it created.

But we can be, because we have inherited a will divided against itself, a division that emerged as the rightful punishment for the free act of rebellion by Adam. That free rebellion upset the order of the world and polluted future acts of will.

Augustine knows that his solution carries him to the edge of a Manichean view; it seems to condense the Manichean view of two contending, divine forces in the world *into* two contending wills lodged in the interior of the soul. And indeed Pelagians, whom he attacks as heretics also, do accuse him of being a secret Manichean. To Pelagians, he has weakened freedom of the will too radically. So, in the very next paragraph, Augustine strives to convert the risk and charge against him of Manicheanism into a strength. "Let them perish from before your face, O God, even as vain talkers and vain seducers of men's minds perish who detect in the act of deliberation two wills at work, and then assert that in us there are two natures of two minds, one good, the other evil. They themselves are truly evil when they think such evil things."[24]

"Let them perish," "vain talkers," "vain seducers," "truly evil." Augustine carries his doctrine to the edge of Manicheanism and then protects it from that reduction, first, by heaping calumny upon Manicheans and calling for them to be exiled to a burning hell; second, by *dramatizing* the difference between two divine *forces* in the world and two *wills* divided against each other in the soul of human beings. Perhaps it is now easier to understand why Augustine said the church actually needed heretics to clarify its own doctrine.

How, more closely, did a will divided against itself come into the world? We draw upon book 12 of *The City of God* to answer. In doing so we also keep in mind how some pagan traditions encountered in previous chapters have themselves encountered divisions within the self. Creon, for instance, struggled with two tendencies in himself—a desire to be a consummate sovereign and a desire to listen to subordinates—but he waited too long in *Antigone* before opting for the second, more promising action. Ismene was divided by a drive to be loyal to her sister and a desire to obey her sovereign. Did she change the initial priority between these two choices later? Some interpreters think so.[25] Augustine's divided will is grounded differently, however, and it carries different existential weight. But the debate between him and them is not between one side denying agency and the other asserting it. It is about the place of gods or god in and after willful acts are taken.

The Vexed Theology of Free Will

Let's now outline the distinctive Augustinian theology of free will along with its solution to the problem of evil, each step in this series setting a pretext for the next.

First, as already noted, Augustine confesses into the interior of himself and a burgeoning Christian culture a god who is creative, omnipotent, benevolent, and salvational, a god who sends those who receive its mysterious grace to heaven and those who do not to eternal hell. The latter burn in hell forever because the divine being restores each body eternally every time it burns into nonbeing.

But, to be more precise, what converts these specific onto-theological *parameters* into a *faith* that people fervently embrace? The Augustinian promise that these parameters provide the highest hope to secure eternal life is one of the lures that converts them from a speculative system into a creed confessed into being. Another is the hope that after the birth of Christ human history becomes providential. From this point on everything said about the problem of evil *must* fit these parameters and the lures that sustain them.

Second, god creates Adam from clay, investing him from the moment of creation with free will. At inception Adam's will is pure; he is capable of willing the good fully on his own cognizance. Much of this story is rehearsed in book 12 of *The City of God* where Augustine presents an authoritative reading of Genesis. There are, of course, other readings of Genesis, including those offered by Harold Bloom after he and David Rosenberg presented a new translation of the first version of the Genesis story, *The Book of J.*[26] Yahweh, as presented there, is a more capricious, blunt, and limited god. Such a reading of Genesis would be anathema to Augustine.

Third, the serpent in the garden is, to Augustine, Lucifer in disguise. Lucifer tempts Eve, who has been created from one of Adam's ribs, with a sinful desire; she in turn tempts Adam. Adam nonetheless assumes all free responsibility for the act he undertakes, eating of the tree of good and evil. For he consented freely to the tempting lures offered by Lucifer and Eve.

Fourth, Adam's act of rebellion against his god is a pure act of free will, not at all impaired, say, by the fact that he is young or inexperienced. This is so because the god had issued a clear command easy to obey.

Fifth, after the free act of rebellion, the god punishes Adam, Eve, the snake, and their progeny severely for free disobedience and rebellion. So, this story of free will is entangled from the start with a demand for obedience to the divine. Now humans lose automatic participation in eternal life; men are doomed to labor in the fields; women are compelled to suffer during child labor; and the snake becomes a humble beast crawling along under the feet of humans. The snake forfeits participation in wisdom with which alternative onto-theological stories in the region had invested it.

Sixth, after this ugly act of free rebellion, all humans thereafter inherit the legacy of a will divided against itself, or in the language used in the *Confessions*, a *double* will. There is no longer purity of free will upon human inheritance of original sin. That first act, and historical human inheritance of it, now make humans responsible for evil in the world. It also creates an existential screen that protects the god from any responsibility for evil in the world he created from scratch. Now divine creation, free will, and human responsibility for evil become enmeshed.

Seventh, after the fall, human beings can no longer positively will the good alone; they can do so only to the degree their will becomes infused with the unfathomable *grace* of god. This outcome, deeply at odds with findings of the Pelagians who had thought humans can will the good alone and earn eternal salvation on their own mettle, preserves the importance of divine grace in the world; it also increases the divine debt to and dependence of all human beings upon god. For now to act freely in support of the good is to receive an infusion of divine grace into the will. The necessity of grace, imputed by Augustine, renders god crucial to life at every moment. Prayer is the way to appreciate that dependence and to receive grace, should it be offered.

The Pelagian creed, Augustine contends, would have eventually drawn more and more people away from attachment to the indispensability and power of god's grace. That is why he worked so hard to get the church to define Pelagianism as a demonic heresy. Calvin will later radicalize the Augustinian doctrine of grace and predestination further.

So we now have an Augustinian doctrine in which a singular god creates the first simple beings followed by a series of fateful events: the intervention of Lucifer, the freely undertaken original sin of rebellion by Adam, a will henceforth divided against itself, and the need thereafter for divine grace become woven together. To remove any of these key pivots would be to render the others shakier. It would also risk burning in hell forever. That is why pagans, Jews, Arians, Manicheans, and Pelagians each must be subjected to severe critique and rebuke in turn. Each jeopardizes, in one way or another, the security and imperial hegemony of the official faith.

From inside the faith Augustine authoritatively confesses a free will that raises humans far above animals; it thereby anchors human exceptionalism, or anthropocentrism, in a divine matrix. For those other beings lack both wills and souls. Free will also offers one explanation of why people are often poised in uncertainty between alternative courses of action, since after the fall all humans inherit a will divided against itself. We, in effect, have willed our own fate. This solution protects the parameters of divine omnipotence and omniscience within which Augustine defined and solved the problem of evil. And it shows why all human beings (after that first pair) *begin* life in primordial debt to divinity.

The Augustinian solution to the problem of evil presses you and me to look first inside ourselves when terrible things happen. How does that work during the Anthropocene when turbulent relations between nature and culture have become all too palpable again, as they had been to so many in the early Greek and Roman days? It is wise to remember that both within the church and outside it there were several noble challenges to the hegemony of Augustinianism. Were there, within those defeated doctrines, themes and/or loose ends that might rise again in new ways?

One loose end floating around in the Augustinian system involves the precise relation of divine grace to free will. If you turn toward god, is that because you are already filled with grace or because you freely turn in the hope of receiving it? That loose end returns to haunt Christianity in the future. Some will push a doctrine of strict predestination; others leave space for free will. The issue returns during the Christian/imperial encounter with pagans in South America and again when white Christians in northern Europe become attached to the advance of capitalist abundance and growth. We will encounter both instances.

Dominion and the Singularity of Time

Let's pick up two more salient themes confessed by Augustine. The dominion of human beings over the rivers, oceans, lakes, lands, animals, and fauna that populate the earth. And the human experience of time in a world governed by a timeless god.

Augustine says in the *Confessions* that only god has dominion over the entire cosmos, even over the experience of time, which he created from scratch *when* he created humans. Here dominion means absolute command, subject to divine will. Later, medieval nominalists will insist that to truly appreciate divine omnipotence and absolute dominion is to see that god is not necessarily subject to, or entangled in, any long-term *telos* or purpose invested in the temporal advance of human beings. According to the most extreme versions of nominalism, he might shift the direction of climate, turn the shape of the oceans, or eliminate humans from the face of the earth at any time. For reasons unfathomable to us. To deny that he could do so is to subtract omnipotence from god, a serious sin. The nominalist movement becomes one of the ways the prose of the world—a world of things and beings said to be invested with divinely inspired purposes that humans can glimpse—becomes divested of those implicit purposes and converted into a more barren desert of Cartesian objects. Cartesianism, Calvinism, and Newtonian science, diverging from one another in other respects, become complementary carriers of such an emergent doctrine.

When Augustine invests god with dominion over the earth and the cosmos, that judgment is balanced by confidence that it will be exercised with divine care for human beings, particularly those who embrace the true faith. So now Augustine can turn to the more confined human sphere of dominion, a dominion over other things of the earth, a dominion qualified by an obligation to exercise it in ways the god approves. A limited human dominion. Here is what he says:

> Therefore, man, whom you have made to your image, did not receive dominion over the lights of heaven, or over that hidden heaven itself, or over day and night, which you called before the foundation of heaven, or over the gathering of the waters, which is the sea, but he has received dominion over the fishes of the sea, and the fowls of the air, and all beasts, and the whole earth, and all creeping things that creep upon the earth. For he judges and approves what he finds right, and he disapproves what he finds wrong, whether in that sacramental administration whereby those men are initiated whom your mercy searches out in many waters; or in that wherein that Fish is set forth, which, having been taken out of the deep, the devout earth feeds upon; or in the signs and utterances of words, made subject to the authority of your book, like birds flying under the firmament, by interpreting, expounding, discoursing, disputing, blessing, or praying to you . . . to the end that people may answer, "Amen."[27]

A set of undulating sentences, replete with adoration of the divine, human authority over animals and plants, and, above all, confidence in the relative stability of those aspects of the world over which God exercises dominion and the abundance of those lower beings over which humans exercise dominion. There had long ago been cataclysmic acts of divine wrath, expressed in the Noah holocaust and, later, the devastating rains and Red Sea drownings when the pharaoh of Egypt persecuted Jews and challenged the authority of Yahweh. But the Augustinian world is now more stable and bountiful.

Dominion changes its meaning in these undulating sentences, as its locus shifts from god to humans, with the first being absolute and the latter being limited to a confined sphere. God has dominion over time, the oceans, and the whole firmament, helping us to see why many evangelicals and conservative Catholics much later came to insist that human beings could not have been the source of things like radical climate change, a slowing of the ocean conveyor, enlarged drought zones and wildfires, the speeding up of glacier flows, or the intensification of hurricanes. Such events, apparently, stretch beyond human efficacies.

It is sinful to even think we could be powerful enough to commit such sins. Human beings, however, because we alone have free will and souls, because we alone are susceptible to grace, and because we alone receive the possibility of eternal life, do exercise dominion over much on the earth. Its fruits are to serve the needs of the most exceptional and gifted beings on the face of the earth. And yet those needs and uses are not to expand so far that the earth itself becomes befouled.

This threefold combination of human exceptionalism, dominion over the earth, and obligations to god becomes rattled again, first, with the rise of European colonialism over pagans elsewhere and, second, with the emergence of the Anthropocene. Was it a mistake all along to invest so much dominion in human agency? Do not other beings on the face of the earth—bats, snails, cougars, bacteria, viruses, crocodiles, coral, forests—also express variable degrees and modes of agency? Is a pandemic the result, then, of a sin against god, as Augustine would think? Or is it the result of species crossings in which bats, viruses, human receptors, trains, planes, public authorities, theo-imbued judgments, and capitalist profit are all involved? How secure is human dominion? How wise is it to act as if the gracious exceptionalism upon which dominion rests provides a solid basis upon which to think, act, and politic?

A threefold formula: god presides over it all, we are sinners nonetheless made in his image, and we are given dominion over the fruits of the earth. As providential history accelerates, after the birth and increasing awareness of Jesus Christ, new catastrophic events—say, a volcano, a massive earthquake, a long drought—must be viewed as divine punishments for deviating from the true path of history. That story line persists for a long time, as is shown by the Lisbon earthquake of 1755 when John Wesley, the famous protestant theologian, blamed the imperial powers of Lisbon for inciting god's wrath and Lisbon Jesuits treated it as the first sign of the second coming, which would occur the next year on that very day.

We must note, then, how impressive and courageous it is for Pope Francis today, steeped in traditions and assumptions of his church as it has evolved over two thousand years, to revise and qualify some Augustinian assumptions. Listen to a few of the points he makes in *Laudato Si': On Care for Our Common Home*:

> 81. Human beings, even if we postulate a process of evolution, also possess a uniqueness which cannot be fully explained by the evolution of other open systems. . . . Yet it would also be mistaken to view other living beings as mere objects subjected to arbitrary human domination.
>
> 90. This is not to put all living beings on the same level nor to deprive human beings of their unique worth and the tremendous responsibility

> it entails. Nor does it imply a divinization of the earth which would prevent us from . . . protecting it in its fragility.
>
> 116. . . . Often, what was handed on was a Promethean vision of mastery over the world. . . . Instead, our "dominion" over the universe should be understood more properly in the sense of responsible stewardship.
>
> 161. Doomsday predictions can no longer be met with irony or disdain. . . . The pace of consumption, waste and environmental change has so stretched the planet's capacity that our contemporary lifestyle, unsustainable as it is, can only precipitate catastrophes, such as those which even now periodically occur in different areas of the world.[28]

In these statements I read Francis to retain papal connections to the Augustinian tradition while—in the face of accelerating and regionally asymmetric climate catastrophes initiated largely by old capitalist states in temperate zones—inflecting that tradition in distinctive ways. It is a delicate series of lines to thread. He is, certainly, not the only one who needs to address the issue. Defenders of Augustinianism, the classical enlightenment, Kantianism, protestant evangelicalism, rapid capitalist growth, productivist socialism, democracy, Anthropocentrism, planetary gradualism, geopolitics, and imperial white superiority reflect a series of intertwined traditions in urgent need of doing so too. I doubt, too, that Francis could write about "pagans" with the disdain that Augustine marshaled in his day. The later history of Euro-holocausts against pagans makes such a venture unacceptable. Francis launches his task in a way that encourages him to call for radical changes in production and consumption while forging spiritual and political assemblages with those in multiple regions attached to diverse theistic and nontheistic perspectives.

The Augustinian discussions of original sin, free will, grace, and eternal salvation rest upon an assumption of eminent causality, a concept, again, conveying the judgment that an entity of high complexity can only be the product of one of yet higher complexity. This assumption/existential demand is one major reason Augustine thinks so ill of nonplatonic pagans. Some—Epicurus, Prodicus, Lucretius, Thucydides, and a few minor characters in Sophocles's dramas among them—speculate that more complex forms evolve from less complex forms through resonant processes of emergence. The processes of emergence, they say, are not yet fully comprehended. But those who promulgate eminent causality also participate in that condition: they construe eminent causality to be incomprehensible to us but real. Speculations about emergence and evolution constitute two of the key reasons Augustine treats a specific class of pagans as *so* lowly and despicable. Onto-theologies of eminent causality contending against ontologies of immanent, emergent causality. Is the Augustinian commitment to the former so intense and severe, first, because it protects human exceptionalism under the province of his god and, second, because it protects the lure of eternal life to which he is so urgently attached?

The issue of eminent causality remains in play when Augustine tackles the quandary of time. Here, he brilliantly examines human experiences of those complex interfoldings between past, present, and future, setting those experiences in contrast to that of an eternal god who rises, mysteriously, above any experience of temporality. That being, indeed, is the author of time rather than existing within its compass.

Augustine, in the relevant chapter of the *Confessions*, notes that pagans often ask Christians what god was doing before he created the world, thinking that the "before" will catch them in a bind: if they say he was getting ready to create the world they admit that time precedes creation. His humorous response is to suggest that god might have been preparing hell for those who ask such questions. An apparently light joke, but one told by a bishop who carried much power in the Roman regime.

His serious answer is to say that god is the bearer of eternity, a concept that finite beings can barely note though cannot understand. It is the very incomprehensibility of eternity to us that endows the incomprehensible being with infinite dignity, creative power, and absolute claim to our obedience.

> At no time, therefore, did you do nothing, since you made time itself. No times are coeternal with you, because you are permanent. . . . What is time? Who can easily and briefly explain this? . . . Yet I state confidently that I know this: if nothing were passing away, there would be no past time, and if nothing were coming, there would be no future time, and if nothing existed, there would be no present time.[29]

We human beings, then, do experience time. Our being is co-primordial with that experience because we are beings who are born and die. Christians must square that ineliminable experience with faith in an eternal god who rises above time.

Some pagans experience time as the recurrence of long cycles. Hesiod did, for instance. Christians cannot accept that reading, however, because, as they see it, god created time with the very creation of the world; humans became monstrous before Noah redeemed them; and the birth of Jesus inaugurated a new, long, singular progressive era to culminate in the second coming.

But how, more closely, do we experience the relation of past, present, and future, and what must be made of that experience? As Augustine struggles with this issue, some profound formulations issue from his pen, or, as he might say, weave their own way into his confessions:

> It is now plain and clear that neither past nor future are existent, and that it is not properly stated that there are three times, past, present, and future. But perhaps it might properly be said that there are three times, the present of things past, the present of things present, and the present of things future. These three are in the soul, but elsewhere I do not see them: the present of things past is in memory; the present of things present is in intuition; the present of things future is in expectation.[30]

Augustine denies that a single measure of heavenly movement, say the speed of light, provides an adequate rendering of time. In doing so, he challenges in advance reductive readings of time later offered by Descartes, Newton, and Einstein in different ways. The flow of time is experiential to him, embedded in human memory, intuition, and expectation, or it is nothing. However, he does not consider the possibility of *multiple* measures and experiences of time, embedded in such things as glaciers, species evolution, viral crossings, and the ocean conveyor.

Today, in the midst of accelerating climate wreckage moving on diverse fronts, we may need to ponder again the issue explored by Augustine; we may be pressed to rethink *both* cyclical and linear, singular images of time, while drawing selective sustenance from both. Note that for Augustine, and for many theologians and secularists who follow him in this respect, it is *human* experiences of past, present, and future that provide the base from which philosophies of linear and/or progressive time are constructed. Many theorists have focused on differences between theological and humanist images of time within that frame, broadly conceived. Those are useful comparisons. But what if today, given contemporary encounters with viruses, bacteria, ocean conveyors, glacier accelerations, ocean acidification, desert expansions, heat domes, and monsoon interruptions, it is wise to address *multiple temporalities of different sorts that periodically intersect?* What if it is timely to ask whether time is composed of multiple temporalities moving at different speeds and different vectors that periodically intersect? As when a virus crosses from a bat into a pangolin and from the latter into humans. Or when climate warming desiccates corals and the coral decline contributes to decline of plankton, fish, and other sea life set on different vectors. What if to proceed beyond Anthropocentrism is to proceed beyond every singular image of time? Trying to think time itself as a composite of bumpy temporalities of different sorts that sometimes intersect.

That is a difficult, demanding question. The way it is posed here will be resisted by Augustinians from one side and Einsteinians from another. Even to pose the question of time is to provoke anxiety, then, sometimes even an existential crisis. For many lives are pegged upon either faith in a progressive march of Christian time or humanistic, secular faith in the singularity of time that enables the steady growth of definitive, scientific explanation. Such assumptions are invested in our languages and other institutions. Neoliberal capitalism exudes confidence that its model of growth and abundance through private profit, military might, and mastery over nature will continue into the indefinite future. To rattle either the Christian or the neoliberal orientation—let alone the contemporary alliances between them—is to sow discord in these fundamental orientations and to project fragility into the institutional trajectories they celebrate.

To pose the issue of time today, then, is to sow civil discord. But to remain within either of the two modern orientations is to court discord from another direction: from destructive forces of climate wreckage that recoil back upon the spiritual and secular institutions that have brought them into being. To ignore this issue today, I wager, is to place old capitalist states, the regions they act upon, and interspecies entanglements at growing existential risk.

In the temperate zones of Europe and several predominantly "settler societies" outside Europe, the Augustinian legacy remains obdurate and intense.

Promising Threads in the Augustinian Legacy

I have been hard on Augustine in this presentation, partly because of the authoritarian ways the bishop of Hippo treated pagans, women, Jews, and a flurry of heretics. And because, as we shall later address, the first Christian/imperial conquest of paganism was later repeated in the Americas, Africa, and Asia. Repeated in atrocious ways that were often worse than the founding version. And the first version helped to set a model that was followed elsewhere and morphed into new endeavors of its own. We will turn to that issue shortly.

But amid such closures within Augustinianism, are there gems that might be pulled out, polished, and inserted into more generous ways of being? I think so. Here are a few.

Augustine confesses an omnipotent god who is the highest object of his faith. He has faith that by doing so he opens himself to the incomprehensible *grace* of that god if it should be delivered. One can respect such a mode of confession while calling upon confessors to *also* confess into being presumptive respect for those who confess other finalities. Two points here. First, by confessing your faith to others you infuse it more deeply into yourself. This is a profound Augustinian insight, an insight about the use of language itself. Words and sentences do not merely describe or designate a world; they help to incorporate a creed into being as they do so. But, second, in confessing a god into being, you also reveal that faith in its truth is not grounded in a set of ironclad, irrefutable arguments.

Closely connected to those first insights is the idea that a particular confessional mode becomes shaken whenever you find yourself moving away from a faith previously imbued in you toward another. As you undergo the *pathos of conversion*. A pathway to possible conversion is littered with innumerable doubts, uncertainties, and anxieties. You become even more a problem to yourself, as Augustine would say. You had encountered unexpected events, say, the delay of a second coming, the dawning of a new sexual orientation, the surprise of a devastating earthquake, the rising of oceans you took to be fixed. They press you to work anew upon yourself and embedded assumptions in your previous orientations to the world.

Out of such an arduous process, a conversion may bubble forth, a new solidification that, after a period of anxiety and uncertainty, replaces your erstwhile faith with a new perspective. As happened to Augustine in his conversion from Epicureanism to Christianity and, again, from Manicheanism to the Trinity. It is unlikely that the new orientation will be as securely grounded in evidence, undiluted revelation, and reasoning as you hoped it would be. But it does become consolidated. And after consolidation, it hovers in the background of things you say and do. The confessional element remains intact, while deniable glimmers of doubt and uncertainty inhabit the new consolidation. Such is the way of being in the world.

Augustine was a genius of the confessional mode, both in the way he practiced it and in the way he commended it as a practice. While deep pluralists must modify some modalities of that practice, we would be unwise to eliminate the confessional mode altogether. For an element of confession attaches itself to most sentences you utter. To articulate one faith, for instance, while seeking to fold presumptive respect for others into it, is to fold the confessional mode into life.

Since we all have grown up with some preliminary faith lodged in us—often replete with ambivalence and doubt—since we today rub shoulders with people of diverse faiths and philosophies, since we periodically encounter unexpected events that may disrupt this or that faith, it is likely that most of us will encounter pressures to undergo conversion to a new orientation at some time or other. The confession/conversion dynamic is not unique to Augustine, then. It is not even unique to those who confess a faith in this or that divinity. Secularists, scientists, and nontheists of various sorts participate in it too.

To confess in such a world may be to encounter traces of doubt and uncertainty in that confessed. Because these two dimensions—confession and doubt—hover together in the same practice. Augustine deployed this combination to express doubt in one round and deepen his faith in the next. We might also deploy it, on occasion, to open ourselves to previously unheard and demeaned voices in and around us. The Augustinian confessional mode, on this reading, thus solicits revision but not elimination.

Another zone of fecund influence is the Augustinian doctrine of divinely endowed human "dominion" over nature. It did not mean complete dominance in the interests of human mastery. Though, under the yokes of exceptionalism, empire, racism, and capitalism, the doctrine has often morphed in those directions. Now that many of us appreciate again pagan ideas of a nonhuman world populated with diverse agencies and forces of multiple sorts that periodically intersect, it may be timely both to appreciate the notion of dominion and to rework it. As Pope Francis, for instance, has begun to do within the church. Dominion now morphs into human tending to the earth and atmosphere, acutely attentive to the diverse agencies and forces that populate them, appreciative of the multiple ways we are entangled with them.

Augustine also addresses paradoxes involved in the organization of earthly governance over a territorial populace. How do you get a populace stuck on the same territory to develop common virtues? How to instill those virtues without already having them to draw upon in doing so? Hobbes, Rousseau, Kant, Hegel, Rawls, and Arendt all explored versions of this paradox and promoted resolutions of it. So did Augustine. To Augustine, only after a territorial populace falls under the thrall of the true god, only after most members confess to that god, does it become feasible to forge a territorial unity under a monarch that both minimizes earthly violence and prepares the way for the possibility of eternal life. Today, many think that his solution requires more violence than it resolves in a world in which people of diverse faiths again rub shoulders on the same territory and enter into close relations with those on other territories. Many of us no longer accept the Augustinian solution. But he saw the problem, and it is difficult, to say the least, to enact adequate responses to it. The paradox of governance persists.

Finally, Augustine on time and the inability to participate in the world without being inhabited by assumptions about that fraught topic. His exploration of the interfoldings of human experiences of past, present, and future attained a high pitch that few in the west have surpassed. As we find ourselves pressed again to rethink time during an unsettling era of climate wreckage, Augustine's efforts provide one valuable place to start. Though they may not be the place to finish.

Are those of us in luxurious, predominantly Christian, capitalist states tethered to institutions projecting an image of time that is increasingly difficult to believe? Can we neither believe in the institutional projections of the future imbued in us nor readily conceive of ways to transcend them? Many respond to such binds with modes of denialism—denial of climate change, denial of dependence upon histories of colonialism and racism, denial of this or that election results, denial of efforts to challenge a world historical itinerary of progress. Rethinking western legacies concerning the shape of time itself must be on the agenda today. Augustine shows us how deep and perplexing the issue is.

And yet, amid appreciation of these Augustinian explorations, it must also not be forgotten how the Augustinian variant of Christianity defined, demeaned, and condemned paganism, how it also defined heretics, Jews, women, gender, and diverse sexualities. In *The Darkening Age: The Christian Destruction of the Classical World*, Catherine Nixey reviews how devastating numerous vigilante actions during this period were against pagans in particular. Time and again, escalating dramatically with the reign of Emperor Theodosius, monks and other Christians went on rampages to destroy pagan temples, kill or maim pagans, burn books, and level magisterial libraries filled with pagan texts. For example, one day in 392 CE

> a large crowd of Christians started to mass outside the temple [of Serapis, a Graeco-Egyptian god in Alexandria], with Theophilus [the bishop of Alexandria] at its head. And then, to the distress of the watching Alexandrians, this crowd surged up the steps, into the sacred precinct and burst into the most beautiful building in the world.
>
> Then they began to destroy it. . . .
>
> As one delighted Christian chronicler put it, the "decrepit dotard" Serapis "was burned to ashes before the eyes of the Alexandria which had worshipped him."[31]

These scenes were replicated often throughout the empire as temples were destroyed, pagan leaders were killed or forced to flee, and stones that had formed the edifices of pagan temples were fashioned into walkways upon which Christians walked triumphantly. The Roman, Christian, Imperial conquest of paganism.

And Augustine? He insistently derided pagan faiths as inferior, sinful, vile, and sacrilegious. He did not call for pagans to be killed, though he did consider riots against them and the destruction of pagan sites of worship to *reveal* the inexorable progress of the world toward Christianity and a final judgment. He often wrote about the suffering of pagans in the passive voice, as if their inferiority meant that they had to give way to the march of progress. These events, while devastating to the parties defiled and crushed, were god-washed from human agencies caught in the putative progress of time. A cover-up of oppression and violence, swallowed into a theology of divine omnipotence and progressive time. The European world would later witness philosophers and theorists such as Locke, Kant, and Tocqueville write in a similar passive voice about the necessary, though perhaps regrettable, demise of two entire continents of pagans in the Americas. Augustine helped to propel that trend, or, as he might say, it launched itself through confessions emanating from him. That is the original sin of Augustinianism, repeated many times after him.

Second Coda

Catherine Keller and Diverse Christianities

Menocchio the Miller

Ovid did not disappear as a subterranean force after the Christian defeat of pagans, though such theo-cosmological speculations usually sank below the gaze of the prelates. One who caught the attention of the authorities was a sixteenth-century literate miller by the name of Menocchio. He had encountered the work of a maverick Augustinian who was equally impressed with Ovid. That priest's account of a world marked by chaos before a god acted upon it mesmerized Menocchio; it fit his peasant experience of how unexpected events occur and worms emerge from rotted cheese. Living in what is now Italy, he loudly proclaimed in pubs that Jesus was sired out of wedlock, that a bubbling world of chaos enabled and limited the work of god, that our bodies rot when we die, and that the promise of an afterlife is a ridiculous, priestly weapon of social control.

The cosmological explorations of this loud rebel, independent thinker, and critic of the clergy, whose circling watermill may have helped to inspire his cosmological thinking, were soon brought to the attention of the Inquisition. The shocking discovery of pagans in the Americas, joined to the rise of Protestantism in northern Europe, had helped to tighten prelate demands to squeeze out any remnants of paganism in the church. Menocchio was subjected to an official interrogation, exquisitely alert to the fact that too many wrong answers in that asymmetrical dialogue could induce the church to burn him at the stake. Here are a few snippets from an early interrogation, as reported by Carlo Ginzburg in *The Cheese and the Worms*:

INQUISITOR: It appears that you contradicted yourself in the previous examinations speaking about God, because one instance you said God was eternal with the chaos, and in another you said that he was made from the chaos: therefore clarify this circumstance and your belief.

MENOCCHIO: My opinion is that God was eternal with chaos, but he did not know himself nor was he alive, but later he became aware of himself, and this is what I mean that he was made from chaos. . . .

INQUISITOR: Did this divine intellect know everything distinctly and in particular in the beginning?

MENOCCHIO: . . . God saw everything, but he did not see all the particular things that were to come. . . .

INQUISITOR: If there had not been that substance . . . , if that chaos had not been there, could God have created the entire apparatus of the world by himself? [Watch out, Menocchio . . .]

MENOCCHIO: I believe that it is impossible to make anything without matter, and even God could not have made anything without matter.[1]

The inquisition continues; but the jig is up. It might remind you a bit of being interrogated by a neoliberal university president who insists he is a CEO and that you are insolent to disagree with him. But things get worse for Menocchio. He was first imprisoned for heresy; he was later released from the dark, dank prison by recanting and promising to refrain from public cosmological speculations in the future; he later yet "backslid," spouting the same theories again; he was finally officially condemned to execution for heresy. As Ginzburg makes clear, something close to these heretical explorations was replicated by several other mavericks at the time, many of whom slipped below the gaze of the prelates or recanted to the dogmatists just in time. For Menocchio himself was a protean pluralist, suggesting that the world should welcome a plurality of final faiths in different regions. The church insisted upon regulating from on high the official theo-cosmology of the day, but it exercised somewhat less effective control over lived cosmologies bubbling up from below.

The Face of the Deep

Augustine issued harsh judgments against pagans. His judgments later carried considerable clout in what became (through conquest) several predominantly white, settler, Christian societies. But, as we have seen, if and when the wraps of institutional authority are loosened, Christianity itself has been, and is, a many-splendored thing, as the multiple differences between Jesus and Paul already revealed in during the early days.[2] Its variations and offshoots often challenge some of the worst aspects of Augustinianism. An engagement with Aquinas would be pertinent here.

Limiting ourselves now, though, to the modern era, there has been, for example, the work of Paul Ricœur on evil, in which he contends that the Augustinian doctrine of original sin was the worst thing that imperial Christianity has foisted upon the world; he then opens himself as a Christian to respectful engagements with a tragic tradition, that is, with one of the pagan traditions Augustine sought to malign and leave behind. In *The Symbolism of Evil*, alert to the Nazi holocaust and the role of imperial Christianity in the indigenous holocaust perpetrated on the Americas, he announces that Christianity must respect the tragic theme that being itself is punctuated by forces not predisposed to human welfare.[3] Even if Christians pursue a more optimistic faith that belies a tragic element, they must, he thinks, engender respect for a tradition that challenges the idea of an omnipotent, benevolent god.

The work of Martin Luther King and Cornel West is important too, as they propel an activist Christianity to fight racism and class inequality. Their works stand as invaluable counterpoints to that of James Baldwin which we have already reviewed, with these three activists eventually welcoming the sort of differences propagated by the others as worthy participants in a society of deep pluralism.[4] Charles Taylor, to take another example, adjusted the Catholic theology in which he had been imbued as he became increasingly aware of its role in purging Amerindians from the Canada in which he was raised.[5] He has demanded respect for indigenous traditions grounded in different experiences of nature, time and human relations to nonhuman inhabitants of the earth. And we have already glimpsed how Pope Francis has recently come to terms with these issues. Finally, it is pertinent to note the work of Mary Daly in rethinking the long patriarchal legacy of the Roman Catholic Church and her pursuit of positive relations between feminism and ecology.[6]

Now, though, I concentrate on the recent work of Catherine Keller, doing so to focus on how one activist, feminist thinker explores the lessons of new geological knowledges for Christian theology and the way minor currents in Christian theology itself can influence them. We gain initial insight into the startling diversity of influences on Keller when we read,

> Just as a convenient example [of the multiple influences on critical theology today], I persisted in theology partly because of early exposure to the work of Paul Tillich, his war-intensified kairos, his God who does not "exist but is the ground of existence." "Being Itself," and his

> inspiration of Mary Daly's verb version, "Be-ing." At almost that same mid-seventies moment, I came into the force field of a Christian deployment of Whitehead's philosophy of process, its take-down of omnipotence, its radical relationalism fomenting John Cobb's early warnings about the global ecology.[7]

Keller will resist a Calvinism which insists that to protect god's omnipotence you must adopt a doctrine of predestination in which everything is decided in advance by god; she embraces, rather, a divinity who cares for humanity and other creatures of god by helping them through hard times, though, as a limited god, it cannot always cut bad things off at the pass before they arrive.[8] And she consistently challenges alpha versions of intellectual masculinity with feminine modes of receptivity and attunement to the world that open recipients to new adventures of thought and action.

We are thus *almost* prepared for the startling move made in *The Face of the Deep: A Theology of Becoming*. Keller challenges the text with which Genesis is traditionally said to begin by church fathers, say, in *The Jerusalem Bible* most congenial to the Augustinian tradition. That biblical version says, "In the Beginning God created the heavens and the earth. Now the earth was a formless void." Such a formulation—even though the small preposition "in" may open a strange door to the god acting in a world that precedes it—otherwise projects an omnipotent, timeless god who created everything from scratch, even time.

Yet, other translations of the Hebrew text, Keller says, open the door to different possibilities. They suggest that the second sentence of the Jerusalem Bible was later inserted into the first, making one long sentence. Take the *Book of J* for example, an attempt by David Rosenberg and Harold Bloom to capture the earliest, the J version, of Genesis. Here the translation is: "Before a plant of the field was in earth, before a grain of the field sprouted—Yahweh had not spilled rain on the earth, nor was there man to work the land—yet from the day Yahweh made earth and sky, a mist from within would rise to moisten the surface."[9] In that sentence it is less clear whether a bubbling world preceded the creation or whether the creation encountered that bubbling world as its precondition.

Perhaps that question—whether there was something there at creation or nothing—was not crucial to earlier carriers of the Hebrew faith. But consider the translation of the same sentence by Rabbi Rashi, an eleventh-century European rabbi who understood Hebrew much better than the fathers of the Catholic Church, and who was writing after several centuries of Christian hegemony and rule over Jews. Here is his translation: "When Elohim began to create heaven and earth—at which time the earth was *tohu vabohu*, darkness was on the face of the deep and the *ruach* was moving upon the face of the waters."[10] Here, again, uncertainty reigns over which occurred first, creation of everything from scratch or a divine creative power that worked upon, and was limited by, a bubbling deep that preceded it, or at least was there when it began the task of creation. The controversial translation of the *Book of J* by David Rosenberg and Harold Bloom pursues a similar aim. The book, attempting to isolate and translate the J version—that is, the first version of Genesis—presents a limited Yahweh who worked upon materials already there and learned from his mistakes as time unfolded.[11]

Keller, a devout theologian, grabs this dilemma by its horns. She offers a translation of the first sentence (or sentences) of Genesis that comes down firmly on the side of a limited, caring god. "When in the beginning Elohim created heaven and earth, the earth was *tohu va bohu*, darkness was upon the face of *tehom*, and the *ruach elohim* vibrating upon the face of the waters."[12] Here *tehom* means depth and chaotic beginning, *tohu*, an unformed condition, and *va bohu*, roughly, echo. This translation, and the interpretation in which it is invested, treats a caring god to work upon a bubbling deep already there. It simultaneously provides the god with materials from which to start and vague limits within which it can act. Moreover, it celebrates a god who lives within time rather than residing on a perch above it. Such a god could even commune with pagan stories of the beginning of things without automatically demoting them to inferior standing.

Keller, in *The Face of the Deep*, draws upon minor chords within Jewish and Christian traditions themselves to save the latter from singular domination by the Constantinian, imperial, universalizing Christianity in which Augustine had invested it authoritatively. She, in effect, also inserts a key pagan element into Christian theology through a textual exegesis that finds it already roaming there. This theological decision simultaneously allows her to challenge the human exceptionalism in which Augustine had set the authoritative church, to open it to explorations of nonhuman agencies within and outside humanity, and, above all, to open reciprocal engagements with pagan cultures that break with imperial/Augustinian practices that had prevailed in Euro-American cultures for centuries. To put it another way, Keller preserves the element of presumptive *care* for the diversity of creatures in the world inspired by an earthy Jesus while challenging closures in the Augustinian *creed* in which that care is contained. In this way, she supports a minor Christian tradition that had, off and on, challenged the major Christian tradition which demanded institutional hegemony in both Catholic and Protestant churches.

There might thereafter be things to draw from Augustine, Luther, and Calvin, but the dominant themes of each required reworking to do so. Reworking so that the reach of divine and human care could be extended together; reworking so that to be Christian does not automatically mean that you have to accept an omnipotent, omniscient god that created both the world and time from scratch. In doing this, Keller also communes with William James, who had earlier embraced the idea of a limited, loving god who faced the rich ambiguity of "litter" in the world, who participated in a world of becoming in which the future was not automatically inscribed in the past.[13]

As Keller puts it, "The action of God is its *relation*—by *feeling and so being felt*, the divine invites the *becoming* of the other; by feeling the becoming of the other, the *divine itself becomes*."[14] That is, this Christian god not only is limited by the materials from which it starts, but also *evolves* in tandem with a world that it assists in becoming otherwise. It does not guide a time it controls to the final judgment. It engages events in the world *within a time that exceeds its reach or consummate control.* The god exceeds us, and it is exceeded by time. It therefore helps humans get through crises that even it does not always anticipate. It is a caring explorer, in communion with multiple other explorers. Not only this, such a god opens us and itself to explorations of other agencies that flow into us and extend beyond our full grasp. Viruses, bacteria, whales, fungi, plants, and so on. It thereby opens us to co-explorations with several pagan traditions—including above all indigenous peoples—who already detect liveliness and modes of agency in the teeming life around them. Here is how Keller puts it in a later book: "That breath of life signifies our interdependence with all that lives in the cycles of oxygen, carbon, nitrogen . . . Here another apophasis comes into play. The myriad other creatures of the earth do not *speak*. They communicate, they signify, they have something like languages; they perform varying degrees of articulation and cognitive sophistication."[15]

It is thus doubly anti-imperial, both in surpassing an imperial god and in pursuing respectful ties across existential differences to non-Christian traditions of the sacred. Those faithful to this image secrete a creed, and that creed is in-formed by a spiritual quest to allow its contestability in the hearts of others to flow into itself as something to respect, doing so to pursue a new kind of pluralism.

There is, then, a human predicament that Keller seeks to face, one that also allows her to enter into appreciative engagements with pagans who sometimes pursue tragic images of possibility and/or often populate the earth with a multiplicity of nonhuman agents with whom they are imbricated. You might say that Keller extends respectful engagements in two directions in *The Face of the Deep:* to both pagan and contemporary minor theologies in the west that themselves challenge major secular and theistic traditions. And, more, to secular theories, to complexity theory, quantum theory, and new work in planetary geology and evolutionary biology that both prod her and provide her with additional resources to mine. And she does mine them. She opens herself, above all, to pagan traditions that invite co-respondences with minor Christianities that locate god within time rather than hoisting "him" above it.

Does Keller pursue a minor Christianity in part out of care for the richness of pagan traditions over which a major Christian tradition has too long claimed unilateral superiority and hegemony? For sure. Is she also freed to loosen up the very tradition in which she was inculcated because she is not now driven by an Augustinian imperative to forge a creed in which the divinity possesses sufficient power—that is, veritable *omnipotence*—to offer the possibility of eternal life to its devotees upon their earthly death? I am not certain, but I do suspect so. She may think that unilateral insistence that other creeds are inferior if they avoid that demand—and if they do not bestow omnipotence on their god to enact it—demands too high an existential price to pay on earth. We have begun to gauge, in this text, the costs such a demand imposes when the creed that promises it is self-bestowed with unfettered superiority over all other creeds. But another point should also be acknowledged. To the extent modern Augustinians embrace the profound contestability of their creed in the hearts and minds of others with whom they rub shoulders, doing so without resenting too much the need to do so, to that extent they too can pursue an existential pluralism that seeks to redress and repair centuries of oppression over pagans, including those who were enslaved, those whose lands were ravaged by settler colonialism, and those who suffered both fates.[16]

A Theology of the Earth

Keller has continued along the trail forged in *The Face of the Deep*. The Euro-American capitalist states that have, to date, released over 50 percent of the cumulative carbon and methane emissions that spawn rapid, accelerating climate warming mostly confess Christianity or a secularism that secretes a rather similar image of time. The Christian right in those states—led by a fascist right in the United States—militantly professes climate denialism. It does so partly on the grounds that an omnipotent god would not allow human beings to change nature or climate on its watch. That denial is joined and amplified by the fact that most white evangelicals and those white Catholics who tend to the right also identify strongly with neoliberal capitalism. The two political drives together—white evangelicalism and neoliberalism—amplify and modulate each other. It is not a coincidence, for instance, that the Republican Party in 2023 could not agree upon a new Speaker of the House until they landed upon a right-wing evangelist who is racist, opposed to gay and transgender rights, a strong denier of human-generated climate destruction, and devoted to ensuring that an evangelical cosmology dominates public life. Racism thus energizes this movement—with the destructive force it propels within and outside Euro-centered regimes. But racism is not the singular source of those energies. It is aligned with other existential sources that help to carry them along. One of them is the cultural imposition of an Augustinian imperial, exceptionalist, insistently universalizing mode of Christianity. Here it is helpful to remember that while the first conquest of paganism often carried racial overtones, the Greeks and northern Europeans conquered were not themselves cast primarily in racial terms.

Keller, in that context, elaborates a feminist variant of Christian doctrine that addresses these issues. Above all, she pursues a mode of spirituality that enacts positive responses to both climate change and the persisting drives to internal and external colonialism that have been prevailed in America. Thus in *Political Theology of the Earth* she pursues a theology that seeks to siphon off the energies of the Christian right and aligns itself with a plurality of secular and nontheistic forces that also take very seriously the climate emergency.

A theology of the earth, then, is one that takes the multiple sources and regionally uneven trajectories of climate destruction seriously, as it also confesses its *own theology* to be contestable—that is, to be worthy of faith for many but not so demonstrable that other indigenous, Buddhist, Hindu, nontheistic, and Christian theo-cosmologies must or should be shuffled beyond the pale of admissibility. Keller is a partisan of one theology who actively seeks alliances and mutual tolerances with those imbued by others, in a world in which multiple creeds persistently rub shoulders within and between regions. She seeks to learn from other creedal practices of the earth while maintaining cultural space for the creed she confesses. Keller is what might be called a deep pluralist, one who aspires to a civilization that houses multiple final faiths and within which the advocates of those faiths evince sufficient existential modesty to pursue a positive ethos of engagements across final faiths.

What, though, is this theology of *the earth*? It is one in which the earth, the planet, is not entirely under the province of a providential god. Nor is it readily susceptible to capitalist mastery. Rather, the earth is marked by periodic unruliness on its own with respect to shifting rates and patterns of glacier flows, species change, drought zones, monsoon cycles, ocean levels, and volcanic eruptions. For example, as a geologist of the sort Keller would respect informs us, "suddenly, a little over 56 million years ago the temperature shot up by 6 degrees Celsius [about 11 degrees F] over a time span as short as

10,000 years, with the poles heating up by 10–20 degrees C over the same period. . . . As in the case of the current warming the ultimate cause of these warm 'spikes' . . . seems to be a sudden rise in the concentration of atmospheric carbon dioxide."[17] All this before human beings made a difference to the ecology of the earth.

So Keller sees that the earth and its atmosphere have a history marked by intense periods of volatility. It is also a planet that responds with exquisite sensitivity to shifting historic modes of intensive human agriculture, deforestation, industrial emissions, and military adventures. Its sensitivity to new perturbations help to explain long periods of slow change punctuated by shorter periods of radical change.

Keller's reading of Genesis, with the god acting upon an unruly deep already there, provides a theological source from which to pursue pluralist formations to fend off racism, colonialism, and climate wreckage at the same time.

It is the spirituality that Keller strives to inspire in us that is crucial. It is a spirituality that defends its creed on one hand while actively seeking positive assemblages with others, in the service of responding to the decimation of the earth that has accompanied both secular capitalist drives and western theologies of divine omnipotence and dominion over the earth. It is also a theology that encourages her to appreciate and listen to vast numbers of creatures and processes that enact modes of agency sometimes difficult to hear, feel, or understand. We now listen to her speak:

> The argument of this book has proceeded from the juxtaposition of the power of the sovereign exception to the potentiality of an ecosocial inception. This contrast registers affectively as the difference between the politics of antagonistic unification and the coalitional capacity of an amorous agonism. . . . Provoked by emergency, the chance may also be growing, against all those odds, of an Eocene emergence: a time of agonized attention to the interdependence of humans with the planetary plenum of nonhumans, not to mention of humans cast as subhumans.[18]

By "amorous agonism" I take Keller to embrace a world in which multiple creedal orientations (religions, philosophies, worldviews, creeds, etc.) speak to one another more closely from reciprocal vantage points of respect joined to the renewed readiness of many to hear murmurings of the earth that had previously escaped them. To hear those murmurings during a period in which the providential and mastery orientations of the recent past advanced by so many have run up against a wall. It is not, for her I think, that all traces of providence have fled. It is, rather, that a limited god needs massive cooperation from human assemblages to accomplish these herculean tasks. In this way we can feel resemblances between Keller, who hearkens from protestant beginnings, and the Pope Francis we consulted in chapter 2 who seeks to adjust Catholic faith to face a new world.

By "against all those odds" I take Keller to appreciate how probabilities of the day press against the new alliances she pursues to respond to unequally distributed global wreckage, while insisting that because the stakes are so high the possibility of positive responses must nonetheless be pursued. It is a theology of the earth listening to a god who cares, without investing that god with omnipotence, omniscience, or, perhaps, the promise of a human afterlife.

Here is another way Keller puts these points, honoring Donna Haraway in the phrases she adopts: "Staying with the trouble means letting go both the chilling optimism of any technofix and its shadow opposite, the surrender to all the too-lates."[19] I seek to commune with and appreciate the *spirituality* of such a theology without embracing myself everything in the *creed* it secretes.[20] That is okay. Pluralist social movements today depend upon diverse constituencies who locate and affirm affinities of spirituality across differences in creed.

3

Todorov, the Second Conquest, and Aztec Cosmology

The Columbus Landing

It is true, as I sang in elementary school: in 1492 Columbus sailed the ocean blue. But he did not discover America. First, he sought to land in India (hence the generic name "Indian" he bestowed on millions of native peoples); and he missed his target by thousands of miles, landing initially in what is today the Bahamas. Second, the land of the Americas was populated by millions of people with rich cultures. Estimates of the preconquest population for both continents range from a high of between 90 and 112 million to a low of 46 to 54 million.[1] Tzvetan Todorov, whose explosive account in *The Conquest of America* of the decimation of Caribbean and South American peoples we will discuss, assessed the world population to be 460 million. The numbers are difficult to establish, and they may not be the most important issue.

Columbus, Todorov says, arrived imbued with high navigational skills, deeply attached to a Christian finalism not that far from Augustine's, a desire to find gold, and, soon, an intense aspiration to give these lands to the king of Spain and Christianity. He returned several times.

While his navigational skills were supple—he became convinced that there was a large mainland close to the islands because of the high volume of fresh water flowing into the sea—his skills of human cultural interpretation were not subtle. Nor did he think they needed to be, since he served the true king and true god. He was thus ruthless in the service of his three aims (gold, god, and king). Whenever he discovered a new island he would give it a Spanish name, treating the name as a mark of its new place under the rule of Spain.

His navigational and naming skills were placed in the service of a finalist reading of history: these lands properly belonged to Spain and the true god worshipped there. Those two convictions and demands are, in the words of Todorov, "always anterior to the experience."[2] We can't entirely fault Columbus for that. There is no such thing as bare experience, despite what a few positivists try to suggest. Complex preorientations are essential to experience itself. Sometimes a new shock or event jolts some to place a few preorientations under critical investigation. As we will see happens to varying degrees to some of the priests who follow Columbus to the "New World." But such an exploratory response does not fit Columbus's character.

His shifts, indeed, follow only a simple pattern: he starts by finding the Caribbean people to be beautiful and cooperative and ends by judging them to be wicked and deserving of every violence he enacted. There was no in-between for Columbus, say, learning from the people you have encountered how they experience life, nature, god, and history, and asking yourself whether experiences you have heretofore placed in those four crude containers might now deserve to be shaken up or modified. Columbus resisted existential disorientation, or perhaps we can say that the numerous shocks he encountered were regularly translated into intensifications of his initial orientations; he adopted increasingly violent strategies to protect them. Anxiety translated into pious aggression. For example, at first he treated the nakedness of young women to manifest the innocence of Eden, later to manifest lewdness and sinfulness. Either/or.

Since his informants called Cuba an island, when he thought it was connected to a mainland, he concluded that these "bestial men" think "the whole world is an island."[3] As a conqueror he brushed over complex syntax to give the objects he discovered new proper names, all the better to claim them for Spain. This was the land of "New Spain." He even changed his name to Cristobal Colón to underline how he brought Christianity to primitive peoples already there, colonizing them for Spain. As Todorov says, "nomination," to him, "is equivalent to taking possession."[4] He carried a mixture of dogmatism, authoritarianism, and condescension to a new world, a combination that continues to find expression in many devotees of conquest. If women were subjugated in patriarchal Spain, it was easy to identify native men as womanly and native women as doubly so. A divinely ordained hierarchy.

So, on the Todorov reading, Columbus oscillated between treating natives as being like him underneath the skin (and therefore assimilable) or unlike him (and therefore requiring conquest). Either/or, with the second view soon acquiring hegemony over the first as new expeditions encountered new resistances. These new pagans even lacked Christian orientations to property.

Columbus was thus a ruthless purveyor of conquest, ready to clear the land of inconvenient peoples to honor his king and his god.

Columbus was also alert to the relations between those external others encountered in a land far from Spain and internal others Christian Spain sought to eliminate from its territory. As he wrote to Isabella: "this present year, 1492, after Your Highnesses have brought to an end the war against the Moors . . . Your Highnesses . . . determined to send me, Cristobal Colón, to the said regions of India. . . . Thus, after having driven all the Jews [and Moors] out of your realm and dominions, Your Highnesses in this same month of January commanded me to set out with a sufficient armada to the said countries of India."[5]

The existential anxieties posed by internal and external others during this period surely resonated back and forth, each intensifying the other. It was the time of the Spanish Inquisition, and Giordano Bruno, a Spanish priest, was publicly burned at the stake in 1600 for advancing a cosmology too distant from the official cosmology of the church and the empire and too close to pagan cosmologies earlier defeated in Europe. I am rather drawn to it.

One thing Todorov misses is how Columbus introduced African slavery into the Caribbean, though he is highly alert to how the relatively immunized Spanish carried the devastations of smallpox onto the Aztec mainland. Columbus soon found the locals to be too susceptible to European diseases to work on the sugar cane he imported. But imported Africans had less susceptibility to those diseases because they had been exposed to them for generations; hence they were also more resistant than Europeans to the malaria and yellow fever that soon accompanied the invasions. Thus the fateful introduction of enslaved African peoples into the islands. Adam Smith had concluded, strictly on instrumental grounds, that slavery was not profitable because of the massive violence needed to sustain it. This early capitalist had not taken into account the role that differential exposure to pathogens has played in the dynamics of colonial history.[6]

Todorov also misses how the devastating Spanish invasion of Mexico helped to intensify the Little Ice Age, a highly variable period that lasted roughly from 1300 to 1850. The best guess is that a combination of reforestation in the Americas after their conquest and population decimation joined a decline in sunspot activity to bolster it. Then a series of planetary self-amplifiers kicked in. Ironically, hardships created by the cold spell in northern Europe probably contributed, along with several other factors, to drive emigration to the Americas. Todorov, alert as we shall see to the effects of plague on Spanish success and Aztec defeat, did not have access to research into the sources and effects of periods of larger planetary volatility.[7]

Augustine, the Christian universalist we have already encountered, was not nearly as ruthless as the devoted invader, Cristobal Colón. Then again, he did not face the shock of a new world totally unexpected by devotees of the authoritative theology he had confessed. If he had faced that shock, would he, as the insistent proponent of Christian superiority over paganism, have become more like Juan Ginés de Sepúlveda (1490–1573), the Spanish priest who construed the Indians to be subhuman, or more like Bartolomé de Las Casas (1484–1536), the priest who first sought to protect them from violence within the orbit of Christian universalism and then stretched that universalism a bit to respond to the strangeness and cruelties of that horrendous encounter? Sepúlveda, you might say, gave priority to Augustinian creed in his ruthless orientation to these new pagans. Las Casas gave priority to the Augustinian ethos of love, more so than his predecessor had done. But that was not enough . . .

Cortés, Sepúlveda, and Las Casas

Hernán Cortés (1484–1547) enters what is now called Mexico in 1519, intent on military conquest. He is a brilliant tactician, effective in drawing non-Aztecs, such as the Toltecs who had been defeated by Aztecs earlier, into alliances with him. He might remind one of fracking companies today moving to a new neighborhood with gaudy promises, only to leave it in a polluted mess once the fracking extraction is complete. The Mexican enemies of the Aztecs, nursing severe grievances against them, did not know what they were getting into. Upon entering a village, Cortés would read the official Spanish statement about mandatory conversion to Christianity and then often exterminate the populace who had just received the ultimatum in a language they did not understand. Todorov even says that he burned his own ships so that his soldiers could not desert to them, though others treat that story as apocryphal.

The conquest of the Aztecs, Todorov contends, was shaped by superior military fire power, by Spanish use of horses, by starvation of the invaded peoples, by the ability of Cortés to make alliances with peoples subjugated by the Aztecs, and, most importantly, by new diseases carried by the Spanish to the Aztecs as they advanced to Mexico City. Was the spread of new diseases part of the strategy of conquest? Well, it might not have been at first, but once the Spanish became alert to how those diseases spread as they advanced, it became so. The Spanish, for instance—seeing how these imported diseases, mostly smallpox, were decimating people who lacked resistance to it—could have concluded that they should depart the region in the face of that knowledge. They did not. Todorov concludes that the population decreased from

60 million to 10 million over a couple of decades. A holocaust brought by Spain to Mexico.

Along the way Todorov asks a question that has since aroused considerable debate. Did Montezuma, the ruler of the Aztecs, fail to put up a major fight because he concluded that these invaders were repeating an earlier invasion that had also succeeded?

I believe that Todorov leans into this idea too strongly. Montezuma, whose military responses *were* rather mysterious, could have fought hard and still retained a cyclical cosmology by surmising that the invaders were repeating an invasion of long ago that failed then and was doomed to fail now. And besides, we now know that Aztecs prophesied that this period—the Fifth Sun—would end through catastrophic earthquakes, not an invasion. Todorov acknowledges that his own conjecture is based on limited evidence. "Unfortunately we lack the documents that might have permitted us to penetrate the mental world of this strange emperor."[8] My conjecture is uncertain too. It, however, is anchored in a comparison between two images of time and more recent evidence drawn from new readings of the Aztecs themselves.[9]

Augustinian cosmology projects an image of time that starts with creation of the world and culminates in a second coming; the human experience of time is presided over by an omnipotent god of consummate power who promises the possibility of eternal life to those who receive its grace. Over and over Augustine chastises pagans of his day because neither their weak gods nor the image of cyclical time they (are said to) embrace can promise eternal life. As the Spanish invaders confronted pagans unexpectedly for a second time, did they now become even more brutal in part to save an image of time that promised historical progress and the prospect of heaven? I don't know, but the suspicion persists. The attribution of cyclical time to the Aztecs (and Montezuma) does not resolve the issue of why they were defeated. But the existential demand to protect an Augustinian image of time may well have folded into multiple other factors that fed the second conquest.

Amid accelerating climate change today the question of time once again bumps onto the agenda, though given the multiple imbrications between the dominant institutions of modernity and a singular, linear image of time, the task is very unsettling. Do perhaps both cyclical and linear images of time need to be reassessed today? Does obdurate cultural resistance to this need feed today into other more obvious sources of resistance to cope with climate wreckage? We will turn to that issue later. Todorov himself embraces an image of linear time not too far from that of Augustine.

While Cortés was the ruthless agent of *conquest* by military means and enslavement, Las Casas was the loving agent of *conversion*. He loved the

Indians, opposed militantly the burnings and killings of Cortés, and sought to convert the Aztecs to Christianity through love. Todorov is clear about how the conquest/conversion pair functioned together to decimate the Aztec world. But Las Casas is nonetheless interesting because eventually, very late in the day, his love of Indians crystallized into a doctrine somewhat less tied to the exclusive universality of Christianity and the Spanish imperative of conversion. Did he risk his own quest for eternal life out of love of Indians?

Sepúlveda was a Spanish priest who drew upon the Renaissance renewal of interest in Aristotle to insist that the Indians were subhuman, incapable of attaining heaven. Their grotesque practices of sacrifice and cannibalism prove them to be lower than human. The only thing to do is to defeat them and to enslave them. He was an ally of Cortés. He and Las Casas thus enter into a long debate.

Sepúlveda, you might say, carries Augustinian antipaganism to new extremes. He is an antipagan on steroids. Las Casas, on the other hand, pushes Augustinian love into new territory. He loves the pagans, but places them under a broadened canopy of Christian universalism. The two priests, through their very debates, expose more dramatically an interdependence and tension that has always been there between the Augustinian *creed* and the Augustinian spirit of *love*, a tension Augustine tried valiantly to resolve. Las Casas, for instance, extends the Augustinian creed a bit to encompass Aztecs in his love. He argues that the Aztecs have an *implicit* image of god that touches Christianity; he also recalls how Christianity itself was founded on the sacrifice of Christ.

Sepúlveda argues that the five plagues sent by Yahweh against the pharaoh set a prelude to the plague brought by conquistadors to Mexico. Both theologians, amid the sharp doctrine/love struggle that absorbs them and wreaks havoc on the Aztecs, continue to attach Christian universalism to progressive history: Noah embodies one threshold in the march of time; the birth of Jesus another; and the Christianization of Rome another. Both priests place these events under the same doctrinal canopy. Meanwhile, Sepúlveda despises under that canopy while Las Casas loves. Amid their struggle, neither was alert to how they pursued two sides of the same coin: conquest and conversion.

Todorov enters the debate at this point to suggest how the postulate of fundamental cultural difference readily slides into the theme of the fundamental inferiority of the other, while that of implicit similarity lends itself to the idea of assimilation and conversion. If you are on the side of the powerful.

He also notes how a culture grounded in sacrifice (Aztecs) must be compared to one grounded in massacre (Imperial Christendom). Can we today, he asks after reviewing the devastation of the Spanish conquest/conversion duo, find a recipe that falls into neither equation?

There may be a potential answer to that question for today, an answer, however, that many constituencies and institutions in the Euro-American world continue to fight against. The multiple sources of that fight deserve reflection. But, first, what would such a contemporary alternative look like?

You could conclude, after reviewing this and numerous other destructive histories with affinities to it, that to live is to be imbued with unconscious precursors to a faith and identity that can easily prompt you to respond to other faiths and identities with disdain, or much worse. You might come to see, too, that few cultures do or can long live in isolation from others, as the Aztec conquests of its competitors even suggests for its day. You could then cultivate, through comparative exploration of diverse cultures and work on the visceral register of life in your own culture, that since the onto-cosmology you embrace has never been proven, and since a strong case can be made that alternative lived creeds are also contestable in the eyes of others, a culture of deep pluralism is highly worthy of pursuit. That would not be a regime in which secular deliberation dominates the public realm and diverse religious faiths are lodged in the private realm. There is no secular space thick enough to handle such a combination; the pretense that there is invariably favors tacitly one faith over others. No, the regime would be one in which several final faiths reach the public realm, and in which participants strive to negotiate a positive ethos of engagement enabling them to seek common solutions to the collective problems that arise.

That is a start, but not yet good enough. The institutions of family rearing, church, school, work, police life, and military life must also become infused with an ethos that encourages connections and alliances across differences in creed. And that ethos? It will be grounded in different ways by different constituencies. Some may ground it in a faith in a generous god. Others in a visceral care for the diversity of life itself, qualified by desires to seek commonalities in new situations.[10]

This will all sound too fragile to many. The argument against that response is that the demand for a universal god, or a universal morality grounded in definitive arguments, or an enclosed white nation, or a secular world in which diverse religious faiths are contained in the private realm, are all even more fragile and more susceptible to violence. To make this counter-argument convincing today, however, it is essential to take concerted actions that simultaneously redress old wrongs against subjugated races and peoples and remove some of the worst insecurities facing white, working-class Euro-Americans so far resistant to such an agenda. Some theorists who support what might be called multifaceted pluralism today fail to acknowledge the importance of this last step. But it cannot be skipped. It at least poses a response to the dilemma Todorov delineates.

Well, Las Casas, the priest who loves the Indians immensely and remains within the creedal orbit of Christian universalism, does make one more adjustment late in life. He had tried to institute a colony based only on conversion, only to see that attempt fail. He now notes how enervated defeated Indian cultures have become. It is as if his love of the other now pushes against the very doctrine in which that love is grounded, encouraging him, amid political risk and existential anxiety, to take a new turn.

Las Casas has argued that Mexicans, like the Spanish, are all "sons of Adam."[11] He finds, late in the day, however, that he now is captured by "a double exteriority." He is not a part of Mexico and has also moved further from belonging to Spain. His critical voice is now ignored on the homeland, though he is spared the Inquisition that Bruno receives. Perhaps because the latter revises the fundamental cosmology of Christianity so much that the earth is no longer placed at the center of things? The cosmos, to Bruno, consists of multiple celestial bodies, with each presenting itself at the center and none residing there. Nonetheless, in his will Las Casas writes, "I believe that because of these impious, criminal and ignominious deeds perpetrated so unjustly, tyrannically and barbarously, God will vent upon Spain His wrath and His fury, for nearly all of Spain has shared in the bloody wealth usurped at the cost of so much ruin and slaughter."[12] He had earlier advocated slavery of Africans though not Amerindians, but that early stance, too, he seems to later have regretted.

The love of Las Casas has begun to push back against the exclusive creed of the universal god with which he had been imbued, even as he hoped for eternal life at the end of his earthly life. The expansion of love makes a difference but its pressure upon the doctrine is insufficient. He thus finds no place to go, either in his thinking or his mode of belonging. A life horrified by all the violences done to an entire civilization whose own sufferings, he would be the first to agree, infinitely outweigh his own. He knows that, at least. His was a futile attempt from within the church and Spain, but at least it was an attempt. Many of us today, in different circumstances, occupy positions somewhat similar to that of Las Casas. That is why the ambiguities of his struggle merit attention, as we struggle with and against the limited options preset for us to appraise. For we, too, are imbued with presumptions at odds with new experiences we encounter. We too may search for constituencies with whom to forge alliances and strategies outside the expected range of preset alternatives. The "we" is hypothetical, prospective, and invitational.

It is perhaps wise to ask what this struggle between Sepúlveda and Las Casas, amid the devastating conquest of Mesoamerican pagans, can teach those of us who inhabit settler societies today. A *creed*—whether a religious faith, a philosophy, or a secular doctrine—stands in an oblique relation to the *spirituality* that permeates it. That relation is one of attenuated interdependence. The first can apply pressure to the other in ways that make a difference. But both, in the life of an individual and a society, must move together for things to change significantly. Creed and spirituality, a twosome that can stray to a degree from one another, also invoke one another. Both are in need of overhaul today, and have been long before today. To grasp this—to feel it—is eventually to call into question closed binary logics that have governed much of modern thought from at least Descartes to today. It calls into question, too, the strange modern image of a world of objects constituted by human subjects. If we inhabit a world of multiple human and nonhuman subjectivities of diverse sorts, then both the monotheistic and secular creeds that have prevailed in the west and the spiritualities associated with them need radical reworking. More about that later.

The conquest/conversion pair, Todorov shows, hid a profound complementarity within its apparent opposition. The two functioned together as powerful strategies to bring the territory of Mexico under Spanish colonization. This complementarity will be repeated in North America, but there—-to streamline things a little—early strategies of conquest through settler invasions were complemented later by a combination of massive military defeats and wrenching drives to spiritual assimilation. In the latter drives, Christianity, English as a universal language, nuclear families, and private parcels of agricultural property became settler demands pursued together. Ned Blackhawk captures a moment in these strategies, as they became consolidated in the 1880s after the second defeat of the Lakota people:

> During the Assimilation Era, U.S. policy makers targeted reservations because they believed that Indian lands needed to be transformed. Only through the Adoption of Christianity, the English language, and Anglo-American practices could Native people become incorporated into the Republic. While railway companies, settlers, and corporate leaders view reservations as sources of unrealized property, reformers targeted Indians in order to transform them. Thomas Morgan of the Commission of Indian Affairs reported in 1890: "The reservation system is an anachronism which has no place in our modern civilization. . . . They should be free to make for themselves homes wherever they will."[13]

As those sentences begin to show, Blackhawk could include white images of time and its image of a world susceptible to human mastery in the intercoded list of imperial onto-civilizational demands. The South American and North American conquests were marked by sharp differences, but that must not allow civilizational affinities between them to be downplayed, particularly during a time of climate wreckage.

Sahagún and the Gods

Gonzalo Guerrero, shipwrecked in 1511 as a youth on the coast of Mexico, went over to the Indians. He taught them about Spanish military strategies. He tattooed his body, married a native princess, wore his hair long and, roughly, became Indian. Did he also internalize the pagan faith and cosmology of the day? Probably, but Todorov does not know for sure. It is clear, however, that this turn by European invaders embodied the best way to escape the odor of Spanish conquest. Secular and religious pressures of the day—bearing down internally and externally from multiple sources—also made such a route extremely uncommon. Nonetheless, perhaps it heralded later modes of syncretism and creolization that carry great promise for today.[14] At that time it merely allowed an escape. It promoted integrity but its singularity did not contribute much to the efficacy of anticolonial efforts. But Guerrero, nonetheless, bore silent witness to resistance to colonialism in one way; Las Casas in another. Even the latter eventually lost his voice in Spain.

Bernardino de Sahagún embodies yet another Spanish priestly response to the pressure cooker of the Spanish invasion. He learns Nahuatl, the local language, better than other priests. His work on Aztec culture poses such a disturbance to Spain that he is prohibited from publishing it. According to Todorov he establishes "an inestimable encyclopedia of the spiritual and material life of the Aztecs before the conquest."[15] Starting by seeking to learn more about an other in need of conversion, he finally discerns how the conquest/conversion syndrome enervates an entire culture without really drawing it into Christianity. He does not assert an implicit identity between cultures that the conquerors refuse to acknowledge (as had Las Casas); rather he strives to read a culture irreducible to the cosmological underpinnings of his own.

But his "distributive" orientation only goes so far. He asserts that Aztec sacrifice is due to profound piety toward the gods, propitiating them because they are so important to life. He sometimes tries to be "neutral" between cultures in the language he uses. But he neither absorbs key aspects of Aztec culture into his own being nor commends a culture of deep pluralism.

In characterizing Sahagún's efforts to record the beliefs and practices of the Aztecs, Todorov says that he often writes in a language that is descriptive rather than evaluative. I doubt that. Different descriptive modes of presentation, you might say, describe from different modes of evaluative or judgmental orientation. Cultural concepts gather together criteria of description from the point of view of specific faiths and cosmological orientations. Such descriptions may express appreciation, condemnation, respect from a distance, or, say, a distancing point of view that releases restraints against violence. But a disengaged point of view is not neutral. The more disengaged a description is the more it relieves worries about dominating, say, relatives, other humans, plants, animals or landscapes involved. It can even provide permission to take charge of the subjects and objects so described.

A distancing, disengaged, objectifying orientation marks one of the dangerous western voices of yesterday and today. It might even remind you of the passive voice Augustine adopted in minimizing the agency of rampaging monks against pagans, locating devastating attacks against pagans within the zone of divine, temporal progress. Here, the agency of the actors diminishes under the divine pull of temporal necessity. Such a passive voice can thus give permission to master, to control, to rule, or even to decimate. It can readily devolve into recipes to dominate nonhuman nature, and, therefore, living nonhuman beings reduced to minor importance within the frame of that image. A secular distancing voice pursues such an agenda in a distinctive key. It drops the will of god and inserts western superiority in a universal march toward material progress.

Later positivist attempts to construct a neutral language have operated as handmaidens of western imperial power. So while a disengaged language might have been the safest for Sahagún to pursue under the watchful eye of the church and empire, that style itself forms a prelude to the Cartesian subject/object duality. It helps vindicate European disengagement from deep concern for colonial and nonhuman worlds, now said to be appropriately susceptible to European mastery.

There is no purely neutral discourse, then. Judgments and preliminary orientations to, say, control or responsiveness or detachment are inscribed in the modes of discourse into which we fall or, on occasion, refine through heightened awareness and concern.

Since we have in fact been engaging Columbus, Cortés, Las Casas, Sepúlveda, Sahagún, and Todorov both to assess their efforts and to prime ourselves to test again received intellectual orientations of the west under new and unexpected conditions, it is pertinent to note what Todorov says about Sahagún's comparisons between Aztec gods and the pagan gods of Europe before the first conquest. One Aztec god is compared to Juno, for example. Here, too, Sahagún chooses to "juxtapose voices rather than to make them interpenetrate."[16]

Yes, but the juxtaposition whereby new Aztec voices are compared to old pagan voices surmounted historically by Christians places both earlier and newly encountered pagans at a severe disadvantage. Intimations of what *we*, those of us who are descendants of settlers, could now learn from such juxtapositions become muffled. The Aztec gods are described by Todorov himself in distancing terms, so the royal "we" cannot learn much positive from them.

The Roman plagues, it has been said, helped to hasten conversion of pagans there to Christianity, since a providential god seemed to offer more earthly protection *and* to promise the possibility of eternal life for the severely afflicted on earth. Did a similar dynamic operate in Mexico? Both the Aztecs and the Spanish tended to attribute greater immunity of the Spanish to the diseases they brought with them to the providence of the Christian god. So, did these differential pathogen efficacies contribute to a conversion dynamic, even if considerable syncretism accompanied it?[17] I do not know.

Todorov and "Critical Humanism"

I admire the way Todorov confronted western radicals and liberals in 1982 with the horrendous holocaust upon which European imperialism rested and later triumphs of capitalist states were built. He confronts Euro-American thought with its self-serving orientations to the other; he makes us feel the incredible cost western colonialism has imposed. It will be said that he does not focus enough on Euro-American capitalism as a catalyst and beneficiary of colonialism, and that is true. Though the reach of his account does pose the question of whether every civilization of productivity—capitalist, socialist, or social democratic—carries within it seeds of imperialism. And he does suggest that no economic system alone—if any such system can indeed be separated as a "sphere" from the larger civilization it invokes and within which it resides—provides the sufficient impetus to imperialism. The Catholic Christianity of Spain folded into the precapitalist economy of the day to fuel an imperial holocaust. Capitalism—giving primacy to private profit over other priorities—is abetted by colonialism rather than being separable from it.

I also admire the way Todorov explores various self-doubts of Christian priests who at first had crossed the vast sea to convert Indians to the universal faith. As they, variously, began to see and feel through horrific experiences how conversion drives feed the impulses to superiority and conquest, several tried to make new adjustments. But, in this or that way, all were caught in binds that compromised those adjustments. Guerrero goes native, but loses critical contact with Spain. Las Casas exudes love, but cannot quite break the Christian universalism and cosmology in which it is set. Sahagún studies Indians closely, but cannot find a way to square that encounter with the Christian universalism to which he desperately clings. Both he and Las Casas, as their thinking and practices evolved, also lost their voices in Spain. Rethinking and reworking thus easily became correlated with irrelevance in Europe. Christian/state imperialism, once it gets going, becomes relentless, though not total in its effects.

I appreciate, too, how Todorov encourages critical intellectuals in settler societies today to draw creative sustenance and self-criticism from his accounts of those priests, doing so to explore binds we must try to negotiate. Is a critique of capitalism really enough? Even though that *is* an essential part of the task. Can we locate internal and external constituencies prepared to contest together colonialisms and the other evils of today? For as Todorov taught, every external other also spawns a series of internal others, and vice versa.

I close this section, however, by contesting another key conclusion Todorov reaches. Doing so to put more pressure on us today. In a later text, *On Human Diversity,* Todorov returns to the question of the other. It cannot be resolved through recourse to a simple universalism, he says, and relativism does not work either. The first, while treating all humans as potentially equal, places cultural historical achievements on a single hierarchy of belonging and achievement. That fosters the conquest/conversion pair.

A relativist ethos, as he reads it, leaves each culture intact, meaning the very worst offenses to human life within a territorial culture are not to be countered. The relativist would simply leave each culture alone, even though that is impossible in a modern world of extensive intertwinements between cultures and diversity within each.

What is left? Well, according to Todorov, "critical humanism." But I doubt that this can suffice. Such humanists place human life at the pinnacle of being. They thus reject, and do not even rework in the light of new experience, Aztec notions of animism and time. They, rather, inject an ontological rupture between human being and the rest of life in order to promote human interconnectedness amid diversity. The world, to such humanists, cannot be a set of nations, each residing in a silo with minimal relations to others. Humanism must guide contacts between nations.

Here are some things Todorov says:

> It is possible to defend a new humanism, provided we are careful to avoid the traps into which the doctrine . . . has sometimes fallen. . . . In this connection . . . it may help to speak of a *critical humanism.*[18]

> The context in which human beings come into the world subjects them to multiple influences. . . . What every human being has in common with all others is the ability to *reject* these determinations. . . . [This is] the distinctive feature of the human species.[19]

> Montesquieu and Rousseau are the ones who refuse to see human life as governed by a seamless determinism.[20]

To overcome isolationism and colonialism, Todorov embraces an inclusive, humanist exceptionalism and, tacitly too, a notion of human culture as internally sufficient to itself in relation to a more or less inert nature that changes only on long, slow time.

Todorov could have modified his inclusive humanism by extracting more insights from Sahagún's engagements with Aztec gods and by examination of the Aztec dicey relations to the cosmos. By "extract" I do not mean that you take these orientations simply as they were—if that were even possible—and then carry them into the different time of today. I mean that as you absorb and ponder them in relation to received Euro-American cosmologies you keep one eye on the American holocaust, another eye on the later, dangerous evolution of capitalist civilizations, and a third eye on the too slow awareness within Euro-America of the accelerating time of climate wreckage it has triggered.

The world is not the oyster of humans (or humanism). It never has been and never will be. Hesiod, Jocasta, Baldwin, Ovid, and the Aztecs understood that better than many imbued with western Christianity and secularism today do. Humanist hopes either to conquer nature or to enter into sweet harmony with it woefully exaggerate the possibilities available. It is better to cultivate a deeper appreciation of the periodic volatility of the earth which has also enabled human and other modes of species life. Tricksters often taught such lessons to northern and southern hemisphere Amerindian cultures.[21] The legacies of Christian providentialism and secular images of mastery, then, together lead western humanists to demand too much from the nonhuman world and to bestow too little vitality upon it. The Aztecs were better on those two fronts, at the very least.

Let me put it this way. In *On Human Diversity* Todorov rightly opposes the kind of scientific reductionism by which human agency is erased and all human behavior is rendered fully susceptible ("in principle") to causal explanation. I concur. But in doing so he confines agency and freedom to human beings alone, creatures said to have broken a determinism otherwise operative in nature. But, no, all kinds of plants, bacteria, viruses, fungi, whales, octopi, ancestors, and jaguars exercise variable modes of agency, whereby thinking and aspiration fold into one another. Entangled as each species is with multiple other agencies and forces, diverse strivings between them sometimes coalesce to bring new things into the world. The 2020 pandemic constitutes one example, as COVID viruses (probably) crossed from bats to pangolins and from pangolins to humans, changing their own composition as they negotiated each new zone of entry. Species evolution itself involves both the strivings of species and confluences with other forces and beings that help to turn it in this or that direction. It does not suffice to say, either, that all beings strive but humans are unique in their ability to think about their own strivings and to adjust them reflexively. Whales do that too; they also join crows, elephants, falcons, and probably many other beings in mourning their dead. Moreover, humans are not that great at it. Human exceptionalism is a dangerous exaggeration, given powerful impetus by western Christian/secular/imperial cultures. Todorov fails miserably here.

It is a further failing of Todorov, closely related to his image of humanity and nature, that he never came to terms with the Anthropocene at all. Even though he wrote well into the middle 2000s. In that, he was joined by a host of humanists, human scientists, physicists, and theologians in western academies. It was possible to launch such an overcoming of western assumptions about cosmology and the planetary. The first overcoming would be to acknowledge the horrendous role of colonialism in creating and sustaining western luxury; the second, to perceive how extractive capitalism, supported by variants of Christian imperialism, has recklessly folded climate wreckage into settler practices of private agriculture, intensive farming, deforestation, soil erosion, cattle methane releases, aquifer destruction, desertification, and so on; until such practices began to place the future of capitalism itself at risk. Today, as we shall see in greater detail in chapter 5, a series of massive capitalist emissions of methane and carbon ignite planetary *amplifiers* that both *augment* the initial effects of those gases and *impersonally distribute* the initial havoc asymmetrically into nontemperate regions that have contributed the least to the carbon emission craze. Today, old traditions of imperialism, besides assuming the old forms, now also ride impersonal planetary circuits of El Niños, trade winds, the polar vortex, glacier melts, desert expansions, weakening of the ocean conveyor, and hurricane intensifications. But we run ahead of ourselves.

Todorov, I fear, also relegated plagues and pandemics to past eras in Europe and Mexico, partly because he was too impressed with modern vaccines and partly because he limited his notion of agency too strictly to humans alone. He underestimated both multiple agencies that roam beyond the human and multiple planetary forces that do so too.

Critical humanism is thus woefully insufficient to the world; it always has been. Today it may be timely for more of us to embrace *multi-entangled humanism* in which dissident citizens in luxurious, temperate zone, capitalist states reach out to dissidents in other regions to foment cross-regional, climate reparative and mitigation movements. Doing so while coming to terms with multiple ways the human estate is profoundly entangled with bacteria in the human gut, viruses that cross from some species to others, processes of horizontal gene transfer that complicate old models of evolution, the long-term dependence of capitalist states upon regions subjected to both direct and impersonal colonization, shifting ocean currents that sometimes aid us and other species and then place both at risk, glacier flows that shift in response to fossil fuel emissions and recoil back upon human life with effects that exceed the triggers launching them, periodic climate changes that alter the conditions of human and nonhuman species all over the world, volcanic eruptions primed to increase in number during the period of climate wreckage. And so on and on. It is not that you give no presumptive priority to human beings if you are an entangled humanist. Or that you deny some special species responsibilities to respond to climate wreckage. It is admirable, for instance, to offer COVID-19 vaccines to as many people and regions that will accept them, even though they kill viruses. And humans in old capitalist states carry special responsibilities to reverse old priorities. It is rather that, first, you acknowledge the complex interdependencies between human beings and nonhuman agencies and forces with which we are insolubly entwined, and, second, you strive to sustain human life in ways that respect the diverse agencies and forces with which they are intimately entangled in a variety of ways.

Todorov was certainly not alone in following such a humanist course. Most Euro-American academicians in the humanities and social scientists joined him through the 1980s and well into the '90s. Many still do. I myself was slow to come to terms with this issue in the early 1980s, and only began to do so as the early '80s rolled into the late '80s. But Todorov tarried way too long, particularly considering the fact that he was in contact with an Aztec cosmology that might have given him some hints. When the delay is that long, then, it becomes imperative to identify assumptions and existential lures that have fostered it. Such engagements may eventually put pressure on you to reconsider prominent doctrines of nature and time that have helped to fabricate the west, as well as the practices of capitalism and Christianity in which those two images are embedded.

Humanists have typically denied relational agency to nonhumans, insisted that the rupture between humans and other species is definitive, adopted the presumption of planetary gradualism, depreciated the shaky place of the human estate on this planet, and insisted upon an image of time as singular and linear. The cumulative result made the imperial west feel good about itself and its colonial practices. But such a constellation also helped to delay western attention to climate wreckage and contributed to failures to probe its distributive effects across different regions.

Another Look at Aztec Gods and Their Cosmos

I would like to attempt a different approach to the Aztec gods, then, however coarsely. I am fortunate to have access to a translation of volume 1 of Sahagún's *A History of Ancient Mexico*. Receiving this text through a series of redactions and translations, I, as amateur and nontheist simultaneously living outside the Catholic Church, the Spanish Empire, and the Aztec world—but within a declining western world—seek to draw selective sustenance from that translation. Doing so during a time of climate wreckage that Todorov himself (and most other westerners of that day) did not recognize, even as it rumbled along. Any judgments extracted from the Sahagún text must be received with doubts and provisos, of course, especially those of an amateur wandering into territory well outside his zone of expertise. Nonetheless, there may be some value in making such an attempt, doing so to learn how to respond better both to past European conquests *and* to current conditions unexpected by children, grandchildren, and great-great-grandchildren of settler agents of conquest and conversion.

I focus on Aztec lived cosmology through descriptions of their gods. The intention is to discern lines of possible affinity (not identity) between European pagan gods and Aztec gods. To do so in ways that can potentially inform modern western life more critically about its own existential demands, lures, and priorities. This is a task that Todorov, as a sympathizer of the Aztecs not himself imbued with pagan sensitivities, does not pursue.

Here, then, are a few Aztec gods. "The god Vitcilupuchrli was another Hercules, of great strength and very bellicose, a great destroyer of towns and killer of people."[22] "Texcatlipoca was considered and held as the true and invisible god who walked all over the heavens and the earth and hell . . . ; he caused wars, enmities and discords."[23] He had some of the qualities of a trickster and adopted various disguises. The jaguar was one of his manifestations on earth. "Tialoctlamaxqui was the god of rain. They said he gave rains to irrigate the earth. . . . It was he who also sent hail and lightning and storms on the water and all angers of rivers and seas."[24] These gods, particularly Texcatlipoca, had trickster qualities and often "mocked" humans who sought to placate them.

"The first one of the principal goddesses embraced by the Mexicans was called Civacoatl, and they say of her that she granted adverse things, such as poverty, mental depression, and sorrows."[25] Sahagún, of course, compares her to Eve after the fall.

"Chalchiuhtlicue was the goddess of water. They honored her because they said she had power over the sea and the rivers, and could drown those who navigated on the waters, causing tempests and whirlwinds that would flood the boats . . . on the water."[26] Now we come to Tlsvulyeutl, "the goddess of carnal matters." (She has different names fitting perhaps different carnal activities.) "It is also said that [she] had the power to produce lust . . . and favored illicit love affairs."[27]

And so it goes, before an even longer list of minor gods is presented.

The Toltecs, adversaries of the Aztecs, also had trickster gods, a god type represented in various ways across numerous Amerindian peoples.[28] "Titlacaoan" for instance, one day "played another trick, disguising himself as a strange Indian. . . . Vemac, (a king) had a very beautiful daughter and as such she was highly coveted by the Toltecs who wanted to marry her [perhaps a euphemism]. He (Vemac) did not wish any of them to have her. She looked toward the market place and she saw Tobeyo [the disguised trickster] completely naked, liked him and on account of the love she felt for him her whole body began to swell."[29] The story continues, doubtless censored in translations we have.

Aztec tricksters, in some ways similar to Dionysus and Proteus, are responsible for disruptions that disturb and unsettle the norms of cultural life, even sometimes bringing them to ruin. They insinuate multiple disruptions into human culture from outside its matrix.

To me, these Aztec gods together present a rather rambunctious cosmos, one that can be rough on human life; the gods preside over tumultuous events and do not reliably place the welfare of Aztecs at the top of their list. They are less beings to worship than perhaps beings to propitiate or even help sometimes to preserve the fifth cosmic order a little longer.

Of course, I can't make too much of my amateurish reading of those gods. But the reading pursued here is re-enforced by the recent work of anthropologist Frances F. Berdan. Berdan says that the Aztec cosmos "had a tumultuous past, an unstable present, and an uncertain future."[30] To her the famed "cyclical" image of time widely attributed to the Aztecs is replete with porosities and ambiguities. Many westerners, to authorize their own image of linear time, streamline the Aztec image too much. Berdan, too, provides a better sense of Aztec worldly experiences that encouraged them to invest so much tumult in their gods—and thus into the cosmos. Between 1460 and 1513 there were at least seven major earthquakes in Mexico. Within roughly the same period there were numerous droughts, floods, and famines.

The Aztec gods challenge the Christian idea of a providential god ruling over a natural history that will be benign as long as you obey it, or in a complementary image, will allow apparently destructive interruptions to propel a benign future in profoundly mysterious ways. It also disturbs secular images of an arrow of time in which (western) humans increasingly preside over a planet which is predisposed to receive that domination. Todorov did not address such a challenge to his own intercoded images of the arrow of time and a relatively benign cosmos.

Berdan also focuses on how Aztecs, with overlapping gods, both worried about how the sun god might end the world soon and practiced human sacrifice to propitiate it and other gods. The epoch they were living in, they projected, would be destroyed by god-induced earthquakes. Yet they complicated the cyclical image of time (that westerners impute to them) enough to strive to convince or help the gods, through ritual sacrifices, to delay that next horrific event in cyclical time. For example, the delicately orchestrated blood sacrifices of enemy warriors were designed to *help* the sun god prolong the Fifth Age by transmitting new energy to it. It was essential to prolong the Fifth Age because when it came to an end humanity would end with it. The cosmos, though, would not end. Another age of a different sort—but perhaps without humans?—would organize itself. Berdan also focuses on how the Spanish built upon previous Aztec cities and temples, reminding those aware of Roman history of how Christian conquerors of European pagans did the same.

In his magisterial book *Mockeries and Metamorphoses of an Aztec God*, Guilhem Olivier adds nuance and detail to these images. At one point he quotes from Motolinía (Toribio de Benavente). Here is what Motolinía says: "when an eclipse, or a great flood, a tempest, earthquakes, epidemics or such things happened that provoked death for everyone or for many, and once that trial or misfortune had come to pass, a new sun and a new age started and indeed they thought that the sun died and that another one was created."[31] Augustine said his god was in charge of the sun. The Aztecs added that theirs could stop it. Today many of us speak of a sun that "dims" at some times, displays sunspot cycles that make a difference to life on earth, and will someday implode.

What can we surmise in a preliminary way about these gods in relation to the human and nonhuman world?

First, by comparison to a Christian single god, they are multiple and often not readily compatible with each other, even though an impersonal ur-god hovers in the background. They enter into messy relations in a messy world.

Second, the cosmos they occupy—including what we call "nature"—is less reliably providential or predisposed to human welfare than either the one historically presented by Christianity or several secularisms that have grown out of it. In some Christian faiths, good events are apt to accompany piety and bad events are divine penalties for bad things done—unless they appear bad to the faithful and are in fact painful moments in the mysterious march of divine progress. In the Aztec and Toltec worlds, many gods did not place the short- or long-term welfare of human beings at the top of their lists. The gods are more capricious, and again from a human standpoint, events are too, partly because they fit roughly into a cyclical world in which life is expunged at the end of one long cycle to be replaced by new modes of being in another.

Third, as already noted, people may not *worship* these unruly gods as much as they *placate* or *help* them, in the hopes of extending the time of the Fifth Age. Sacrifice, in turn, may be aimed at transmitting essential energies to the gods, to help them extend a Fifth Age whose end will also mean the end of the Aztecs. They have no idea what will come after the end of that Age. Sacrifices, enacted in solemn rituals, implant more firmly in the sedimented memories of a highly verbal culture the dicey relations between humans and cosmic god-forces. To help the gods through blood sacrifice, of course, means that they are not omnipotent.

Finally, the gods projected do not occupy time as an arrow—as Christianity and secularism postulate time to be; they, rather, participate as powerful agents hovering over a loosely cyclical world, a world composed of cycles within cycles. A cycle can stop the sun, foment earthquakes, or fashion a massive flood to end an entire era of human life . . . , before a new one starts that may or may not include humans. The ambiguities the Aztecs encountered in maintaining both of those views—powerful gods and cyclical time—may be roughly comparable to those Augustine encountered in fitting human agency and divine grace together in a world inexorably pointed toward a second coming. Perhaps pagan notions of fate are better compared to Augustinian notions of mysterious grace than to his formulations of free will—since free will for him could not be sustained without grace. What, though, about later secular images of time in which grace is denied?

There is another difference, already implied and more subtle in character: Augustinian Christians honor a line of associations flowing through dogma, belief, worship, and obedience, with a specific subjectivity attached to those elements. And Aztecs? They seem more attached to aiding gods through sacrifice than to worship of them, to appreciate habit more than dogmatic belief, and to engage a world not reliably oriented to them through providence. Though each of these three contrasting sets of terms can slide or bleed into the other (worship/help, belief/habit, providence/capriciousness), *the priorities and crystallizations vary significantly between the two cultures*. The distinctions between these two cosmic orientations thus do not correspond to a binary form. Which probably helps to account for the difficulty Jesuit and Dominican priests had in interpreting peoples they held from the start to be inferior.[32]

I have still not penetrated deeply enough into Aztec cosmology, however. James Maffie, in his impressive recent book *Aztec Philosophy: Understanding a World in Motion*, sets the stage to dig more deeply. He argues that the Aztecs embraced a process cosmology. The cosmos—*teotl*—functions as a set of self-composed twosomes that both depend upon one another *and* stand in agonistic relations to each other. The actuality of becoming—the emergence of new beings and things in time—emerges through such imbrications. The world is thus not anchored in a binary or two valued logic; to bring such a logic to it would be to block comprehension of its dynamic processes as well as of the mysteries that arise for Aztecs themselves.

Life and death and male/female sexual intercourse are two of the interwoven, generative processes. Life differs from death, but death is generative of new lives of different sorts. The living, for instance, feed on the carcasses of the dead, and the dead fertilize new life. The rhythmic back and forth of sexual intercourse similarly encodes agonistic twosomeness of tension and passionate intertwining. These examples are real in themselves; they also encode or illustrate how the cosmos itself generates and regenerates. Luckily, the Aztecs do not invoke pool tables or windup watches to serve as models of causality. Nor binary computers.

The gods, at least in Maffie's book, are less actual beings and more human crystallizations of relevant forces that intersect and compose a bumpy cosmos of becoming. But in an email to me Maffie said that he no longer holds that view.[33] The gods are real. That reading helps to explain how blood sacrifice can be so important in transmitting new energies to the gods and how those energies in turn can help to extend the time of the Fifth Sun. I admit that I wish that Maffie had explored how this Aztec process image of a world of becoming—each ending in a specific mode of devastation—compares and contrasts to later western process cosmologies advanced by figures such as Nietzsche, Whitehead, Gilles Deleuze, or Michel Serres. But the themes he does pursue of self-organization through tensions and weaving, of pantheism, and of animism are transfixing. So is that of a different cosmic force invoked at each long interval to bring an entire age down: worldwide floods, fires, volcanoes, and earthquakes among them.

Maffie also briefly notes how Aztecs elaborate "three dimensions of time running concurrently" and how there are "multiple micropolitics of becoming of the cosmos."[34] These brief asides dangle tempting preludes for the image of time as a multiplicity we will consider in the last chapter, during a new time of climate wreckage.

Weaving is one of the defining processes and metaphors in Aztec cosmology. New entities are woven into place; the cosmos itself is composed into five ages through the warp and weft of weaving. Xochiquetzal is a youthful goddess of sexuality, weaving, and generation. "The Aztecs associated her with exuberant and female sexuality, desire, fecundity and generative power, along with flowers, feasting, and pleasure generally."[35] Xochiquetzal may remind you of that weaving contest between Minerva and Arachne in Ovid, as the two competitors hitched up their skirts to compete. Arachne, the winner, was transformed into a spider by the goddess in a way that encourages us to appreciate the rich life, silk release activities, and weaving capacities of spiders. Of course, there is no identity here between these two pagan cultures. There seldom is.

I note, too, a moment in Maffie's account that may suggest the possibility of a close encounter between Jocasta and Aztecs, if the vicissitudes of western imperial history had allowed such an encounter to occur. The precarious world inhabited by humans in the Fifth Age, he says, is marked not only by the uncertainty of when it will self-destruct: the Aztecs have no idea what *teotl* will bring next. The Fifth Age is marked, too, by multiple "crossroads," fateful crossings where people and regimes take one turn rather than another, sometimes wisely and often foolishly. Loose cosmic energies cluster at each crossroad, ready to be mobilized in one direction or another. Are these choices and results preordained? Do they embody contingencies or chance to some degree? Is it possible to become wiser about what to do at such a crossroads? Did all Aztecs agree on answers to these issues? These issues, as they also do in the story of Jocasta, may remain shrouded in uncertainty or mystery. But I do wonder whether both the shamans of that day and Sophocles in another day participate in an incipient image of wisdom here: when things are going well, do not become too hubristic or imperial. A longer bout of questioning together of Jocasta and the Aztecs could well be worthwhile, though the pursuit of *comparisons* between them is inapt to lead to *identities*.

Nepantla processes, says Maffie, "bring, join, unite or interlace together two or more things in a manner that is simultaneously creatively destructive and destructively creative, and therefore transformative."[36] That being said, I now let Maffie himself close this brief foray into his text, as he summarizes nepantla as creative/destructive motion:

> Human existence is, in short, *in* nepantla and *of* nepantla. It is *in* nepantla in the sense that it is defined by ceaseless and ineluctable betwixt-and-betweenness. Life takes time-place in the cyclical back-and-forth struggle of cosmic inamic pairs such as order-disorder, being-nonbeing, life-death, and male-female. Trapped in the tension-ridden, reciprocating balance between being-nonbeing, human existence is defined by inescapable processing, becoming, and transformation. Human life is consequently unstable, fragile, perilous, fleeting, and evanescent.[37]

The nonhuman world, to Aztecs, is thus not reliably tuned toward them as its unique beneficiaries. Indeed, there is neither a fall from complete clarity in a world to which they belong nor a positivist reduction of that world to practices of mastery. Precarity is wired into being itself; the nonhuman world is rocky from the point of view of humanity. And, perhaps, one spiritual task is to fold appreciation of the grandeur of this world more intricately into cultural life.

What, then, if European paganism had continued to evolve for several centuries and then encountered Aztec paganism during the sixteenth century without either culture being subjected to a long period of domination by compulsory Christendom? That is, what if European paganism had evolved openly alongside Christianity *within* Europe before contact with the American pagans had been made? These are the kind of counterfactuals many warn against projecting, since they range far from what actually happened. But every interpretation, recall, spawns such large counterfactuals, whether actively pursued or tacitly adopted. So it may be wise to entertain this one during a time of climate destruction.

It seems probable that the old world and the new world (to Europeans) would still have fought. But would they also have uncovered more affinities through which new creolizations could evolve more widely and rapidly, syntheses that avoided or curtailed the monstrous holocaust occasioned by the Christian/Spanish conquest of Mexico? Would they even have evolved co-understandings that allowed their inheritors to perceive (acknowledge) the wreckages of the Anthropocene at an earlier moment? These are, of course, huge historic counterfactual possibilities to entertain. But, in the light of the two European conquests of paganism, of later Christian/capitalist productions of fascism, and of the Christian/capitalist production of the time of climate wreckage, maybe more people need to pursue threads and possibilities emanating from two pagan eras to help them come to terms with the planetary and existential problems of today. We need to extend explorations of human relations to the cosmos beyond the terms set by dominant, western cosmologies.

You do not need to accept everything gleaned from such excursions into European and Aztec paganism to ponder what insights each could have spurred today if they had continued as recognized cosmologies contending with and against the Christian set. For European and settler societies ushered in horrific conquests, land grabs, intensive agriculture, and capitalism, and then helped spawn the time of climate wreckage. And contemporary inheritors of these former traditions are still with us to help ponder the issues.

Sophocles's messenger, near the end of two of his tragedies, seems both to assume that things can get out of hand when nonhuman events collide with human passions and to call into question familiar readings of those dramas that treat the results as preordained. Then he says of his alternative reading, "make of this what you will." The alternative reading, playing up the role of contingent intersections in human and nonhuman life, is offered as food for thought rather than presented as definitive. That is how I locate this preliminary and amateurish rendering of Aztecs in relation to us today, too.

To invoke both Aztecs and Sophocles is to solicit cultural *interpenetrations* between two cultures that now float outside the mainstream, rather than to make mere *juxtapositions* between them. What selective contributions can two paganisms make to rethinking Christian/capitalist/secular worlds under current conditions of being? One way may be to encourage more of us to fold planetary bumpiness more robustly into our own thought about culture and efforts to address climate wreckage. Another, perhaps, is to explore how to appreciate the grandeur and fecundity of a world that displays such bumpy temporal characteristics.

Every culture evolves. European secularism, for instance, evolved out of Christianity and continues to do so. Capitalism evolved out of feudalism. The recent theme of horizontal gene transfer in species evolution evolved out of the slower-moving tree-like model of Darwin. Quantum theory evolved out of Newtonianism, as it challenged the core principles of the latter. What could have happened if, rather than two conquests, two paganisms had evolved alongside the emergence and evolution of territorial Christianity? Would Euro-American cultures be less destructive, more profoundly pluralistic, and better prepared to grasp and cope with climate wreckage differentially facing the world today?

With two eyes on the time of climate wreckage, is it possible to locate another cosmology somewhere between the cyclical images of the Aztecs, the providential images of Christianity, and the linear, singular images of western secularism? My amateurish instinct suggests that engaging pagan philosophies in old and new worlds can help with such an endeavor, particularly when those engagements also draw upon speculations of dissident moderns touched by these very traditions. When a sperm enters an egg, resonances are set into motion between sperm and egg that sometimes spawn a novel evolutionary result best described through the language of metamorphosis. Similar processes are at work in horizontal gene transfer when viruses or bacteria ingress into an embryo. It also aids our thinking to grapple with a cosmology in which the order of the cosmos itself is not highly or reliably disposed to human welfare. A major earthquake, for instance, could end one era and start another for the Aztecs. Can a preliminary engagement with Aztecs also help prepare the way to better grasp and appreciate the post-pagan cosmology of Michel Serres? We shall see.

Conclusion

Given the boldness and care of his engagements with human others encountered by aggressive Euro/Christian cultures and given the probable fact that Aztecs could have opened other windows, too, for him to ponder the volatile character of human/nonhuman relations, Todorov's critical humanism and planetary gradualism are far too limited in their awareness of planetary volatility. Those who have inherited the presumptions of "the west" must do much better, while continuing to learn from him. Human exceptionalism, planetary gradualism, and linear images of time are simultaneously highly problematic and deeply woven into Christian and post-Christian secular cultures. Neoliberal venture capitalists, for instance, push these themes hard. Such affect-imbued assumptions and practices first distorted Christian engagements with pagans in Europe; they then did so in horrific ways with Aztecs and other Amerindian cultures; and they then blunted, until late in the day, close encounters with the climate wreckage that has been rumbling beneath our feet and above our heads for a long time. Could Todorov have learned more from those Aztec gods about the shakiness of the human place in the cosmos, about impersonal planetary circuits of imperial power, and about the vicissitudes of modern, western images of time? We turn again to those issues—the shakiness of the human estate, impersonal planetary distributors of imperial power, and time as a multiplicity—in the last chapter of this study.

Third Coda

Tocqueville and White Settler Society

In 1958 John F. Kennedy, in a book titled *A Nation of Immigrants,* responded critically to yet another renewal of white, Christian nativism. He did so by insisting that the United States is a country of immigrants of diverse sorts. That theme made sense to many Americans who, like me, were products of recent European immigrants. My father's parents had emigrated from Ireland early in the twentieth century, fleeing "the time of troubles" there. Kennedy wanted to welcome Italians and other southern Europeans, those whom nativists would exclude on the grounds of religion, color, and class. Most Italians and Irish were Catholic, while white nativists were Protestant and could brag of a longer heritage on American soil.

While one side in the recurrent nativist/immigrant debate supports greater diversity than the other, the terms of that debate itself occlude a profound feature of American life. The United States, like Australia, Israel, New Zealand, and regimes in South America, is a settler, colonial society. It is founded on the violent displacement of highly cultured peoples who were already there. Unlike some other settler societies, the United States is also grounded in enslavement of African peoples imported violently and on the conquest of Mexicans who used to control what is now the southwest United States. Three early sins of the United States are thus its violent expulsion of Amerindians from the continent, the early initiation of enslavement upon which much of the wealth of the continent was built, and the conquest of Mexico on the current southwest territory of the United States. Three sins reverberating with each other, the first silenced by immigrant/nativist debates, the second occluded by it, and the third more or less forgotten. There is a fourth sin, too, working in concert with the first three. We will turn to it soon.

Roxanne Dunbar-Ortiz, in two resounding books, traces the violence upon which the United States, as a continental regime grounded singularly in private property, was founded. Whiteness itself, as a generic descriptor, was fabricated through these processes, supplementing or even displacing earlier descriptions of oneself as, say, English, Irish, or Scottish. She is herself the daughter of a part Cherokee mother who grew up in the current state of Oklahoma after her predecessors had been placed on a forced march from Georgia. We will turn soon to her unvarnished histories of enslavement, Amerindian conquests, and the conquest of Mexico. Before that, it may be wise to consult a renowned theorist of early American democracy who smoothed over these processes as they were being launched. That may underline the amount of work Dunbar-Ortiz and others have to do.

Democracy in America

Here we focus on Alexis de Tocqueville. The aristocratic French thinker visited the young American state in the 1830s, celebrating its local politics, its commercial energies, and, less confidently, the narrow balances between diversity and equality it pursued. He also, to his credit, addressed two violent and founding features of the young republic: its enslavement of those wrested violently from Africa and forced to work the cotton, sugar, and tobacco plantations of the South and its historic drives to replace millions of Amerindians as sovereign peoples who ranged across the entire continent.

Tocqueville was against enslavement of Africans on Christian grounds, but once installed he saw no credible way to get out of it in America without posing threats to the advanced civilization under way. His formulations about it are thus stated with caution and infused with regret. They were roughly consonant with positions Jefferson had staked out in a series of letters written between 1793 and 1824. The differences between the races are so severe and immutable, Jefferson had said, that they could not hope to occupy the same territory together as free citizens. The Africans should thus be freed and shipped gradually to another territory, Santa Domingo, if possible.[1]

Slavery, Tocqueville himself insisted, is inconsistent with Christianity, one of the founding sources of his thought and of the new American civilization. But once slavery was installed, as it had been for generations in the United States, dilemmas accompanied attempts to remove it. Why? The two races had different privileges, suffering, and mores folded deeply into their bodies. Tocqueville's insight into mores was indeed acute, though his racialization of them was highly destructive. Through repeated institutional behavior, mores do become installed in common prompts and precursors to cultural life, setting preliminary affect-infused pivots to judgment and action. Disrupting old mores can meet with obstinate, sometimes violent, resistance. To pluralize a culture, it is thus also incumbent to work assiduously on entrenched mores that resist pluralization.

Back to Tocqueville. Southern white aristocrats, he said, had become immured in a way of life dependent upon the enslavement of others; it was encoded in their interests, expectations, and white privileges. This all weighed heavily with Tocqueville, much more than did the immense suffering of the enslaved.

So, the dilemma, as Tocqueville understood it. You could free Negroes and try to live with them on the same territory. But that won't work. Alternately you could try to return them to Guinea, Tocqueville said, a land on the western edge of Africa. But they have become too numerous for that. "Once one admits that whites and emancipated Negroes face each other like two foreign peoples on the same soil, it can easily be understood that there are only two possibilities for the future: the Negroes and the whites must either mingle completely or they must part."[2] The latter is foreclosed, the former is highly improbable.

But why is the former so unlikely? Tocqueville cites the proud spirit of idleness and aristocratic mores among white southern elites; he also senses a rise in racial resentment among poor whites if equality is pursued.

His sense is that an impasse is almost unavoidable: either enslavement persists and erodes the mores of a white, Christian, commercial culture or it is overcome and erodes the unity needed for a vibrant civilizational culture. Blending, the multiplication of diversities on the same territory—what today would be called creolization and was already under way in some Latin regions at the time—cannot succeed, he says. So, shall we say that Tocqueville has identified a tragic impasse, wired from the start into the fateful decision to enslave?

I don't think so. To come to terms with tragic *possibility* can sometimes be to assert that the worst could occur unless social movements succeed in loosening the binds closing in upon a populace. Such a vision solicits action guided by the onto-judgment that no divine providence works independently and reliably in your favor. Tocqueville, however, himself identified a providential tendency of history, as we shall see more closely when we chart his account of the relations between white settlers and Amerindians. He was more prepared to sacrifice territorial enslavement to such a providential history than to free enslaved peoples within the same territory. He thus does not pursue a tragic vision of possibility, of the type delineated earlier in this text by James Baldwin. Tocqueville, the pluralist, could not imagine social movements to forge a pluralistic, creolized culture that breaks radically with the hegemony of the Protestant white culture he encountered.

He thus could regret slavery in ways that relieved his sense of implication in the institutions, practices, and mores that sustain it. While he embraced a Christian idea that all humans are valued in the eyes of god, he also embraced a hierarchy of degrees of humanity. So: Freedom and separation, perhaps; freedom and territorial pluralization, no.

We also begin to see how the white, Christian civilization he embraced, on a territory also populated by enslaved Africans and indigenous peoples, set up a future in which racial and Christian unity would repeatedly be embraced to submerge or overcome critical attention to real class divisions. Whiteness is a politico-cultural formation that helps to align a now whitened lower class with a whitened upper class by enslavement and/or genocide against two other peoples.

When we turn to Tocqueville's engagement with Amerindian peoples thriving on the American continent before Europeans arrived, some of the assumptions stated above stand out more starkly. Tocqueville concludes that Amerindians are already on the way out. The task is to grasp the historical inexorability of their regrettable demise.[3] An early version of replacement theory.

We sample a few sweeping formulations, early in the text, that set the stage for the entire two-volume study of America:

> These vast wildernesses were not completely unvisited by man: for centuries some nomads had lived under the dark forests or on the meadows of the prairies. . . .
>
> . . . The Indians occupied but did not possess the land. It is by agriculture that man wins the soil. . . .
>
> One could still properly call North America an empty continent, a deserted land waiting for inhabitants. . . . In this condition it offers itself not to the isolated, ignorant, and barbarous man of the first ages, but to man who has already mastered the most important secrets of nature, united to his fellows, and taught by the experience of fifty centuries.[4]

"Wildernesses," "nomads," "did not possess the land," "agriculture," "the secrets of nature," "an empty continent." Echoing the Augustinian language of progressive historical fate inexorably facing European pagans who could not keep up with preordained history, the future of these new pagans is inscribed in the very terms through which they are described. They do not divide the land into private parcels and farm the soil intensively. Thus they lag behind the historical march of civilization. Description infused with cultural judgment. They are the walking dead. You may regret their fate, particularly the massive violences through which it is enacted. But you must not try to forestall it. You can, of course, try to protect your moral rectitude by protesting this or that outrage as the inexorable march of a white, Christian civilization proceeds across the continent. Tocqueville infuses race more deeply into the Augustinian story.

One more theme is needed to complete this intensification of the Augustinian imperative on a continent still new to northern Europeans. It is not long in coming. It is assertion of necessity to inscribe *the same cosmic faith over an entire territorial regime* if that territorial civilization is to flourish. A deep plurality of constituencies differing in racial, religious, and cosmological mores as they rub shoulders together on the same territory, is incompatible with the possibilities of high civilization itself. At least with the civilizational imagination of this European.

> In the United States it is not only mores that are controlled by religion, but its sway extends over reason. . . . There are some who profess Christian dogmas because they believe them and others who do so because they are afraid to look as though they do not believe in them. So Christianity reigns without obstacles; as I have said elsewhere everything in the moral field is certain and fixed, although the world of politics seems to be given over to argument and experiment.[5]

So now you have it. Whiteness, Christianity, intensive agriculture, private parcels of land, and common mores imbued into white souls and institutions are woven together into a complex so that each supports the integrity of the others. They together define a distinctive civi-territorial complex. That means, among other things, that Amerindians must continually retreat as white Christian settler farmers expand violently across the continent. It implies also that there is no way to combine on the same continent the communal modes of farming and land management of those already there with a division of land into private parcels owned by white, Christian farmers. For only the latter farm the land efficiently, rather than merely wandering lightly across it.

Tocqueville has constructed a white, Christo-territorial civilization that must swamp everything that preceded it—or even for a time resided alongside it. For Tocqueville, it is fated to dominate the continent. We will soon see how Kant completes this story of intertwinement between Christianity, race, time, commerce, and reason itself, doing so in ways that weave strings of temporal progress into the putative essence of reason. Both Tocqueville and Kant were, I am told, nice people. They also coalesced to plaster a veneer of regrettable necessity over whiteness, empire, racism, Christianity, civilization, and violent displacement of a subjugated people.[6]

A Pequot Reply

William Apess, a Pequot who might have been the first Amerindian to write in English, composed in 1836 an essay on King Philip, the Pequot leader who had lost a valiant battle to protect Pequot territory against English Puritans a couple of centuries before. It can be read as a reply to Tocqueville:

> O Christians, can you answer for those beings that have been destroyed by your hostilities . . . ? And will you presume to say that you are executing the judgments of God by so doing? . . . And as the seed of inequity and prejudice was sown in that day [the early Puritan invasion], so it still remains; and there is a deep-rooted popular opinion in the hearts of many that Indians were made . . . on purpose for destruction, to be driven out by white Christians, and they to take their places; and that God had decreed it all from eternity. . . .
>
> . . . But must I say, and shall I say it, that missionaries have injured us more than they have done us good, by degrading us as a people, in breaking up our governments. . . . Oh, what cursed doctrine is this! . . . But I would suggest one thing, and that is, let the ministers and people use the colored people they have already around them like human beings, before they go to convert any more.[7]

Apess links Black enslavement with depredations against Amerindians, treating both to be effects of the same theo-civi-territorial impulses, ambitions, and implacable demands. His was a voice that had to be marginalized in order for white, settler society to advance further, to construct memories that consolidated it through racialization, and to feel good about itself.[8]

The Violences of White Settler Society

Roxanne Dunbar-Ortiz, in *An Indigenous Peoples' History of the United States* and *Not "A Nation of Immigrants,"* peels off the veneer and rhetoric of regretful civilizational necessity pasted over settler colonization by Tocqueville and many others. She focuses on the violences, horrors, and illegalities. Here we will chart a couple of exemplary moments, moments that capture settler violences grounded in an insistent image of civilizational unity and superiority. We precede them by noting that the European invaders typically thought that latitude and climate varied together, not knowing how climate can change rapidly in ways that break with those assumptions. They were in fact in the middle of a little ice age that was threatening alike to settlers and those whose lands were invaded.

In fact, an authoritative book of the day, published by Richard Hakluyt, a minister and naturalist, ignored this climate change when he encouraged settlers to consider Jamestown in 1607 as the opening of a possible water route to the Pacific, and generally anticipated a "warm but temperate country where the English could grow olives, grapes, citrus, and almonds and raise silkworms."[9] I do not note these mistakes about climate and the size of America to ridicule the writer, since even as late as the 1960s and '70s most western geologists, climatologists, and humanists labored under the assumption that climate and species evolution always change at a very slow pace. We doubtless labor under other illusions today in this or that zone of knowledge. Certainty and immodesty are defeated over and over, only to rise again.

In 1620 after the landing of the *Mayflower*, smallpox spread from the invaders to Pequot fishing and farming communities, for the Tocqueville story of Amerindians as only nomads was a gross exaggeration. Puritans, anchored in a Calvinist theology that echoed and soon racialized antipagan themes from Augustine, now defined nature as lands subject to mastery. That justified the devastating war against the Pequot. William Bradford, a Puritan and governor of Massachusetts, wrote approvingly about the fiery holocaust at the time. "Those that scaped the fire were slain with the sword; some hewed to pieces, others rune through with their rapiers. . . . It was a fearful sight . . . but the victory seemed a sweeter sacrifice, and they gave the prayers thereof to God, who had wrought so wonderfully for them . . . and give them so speedy a victory over so proud and insulting an enemy."[10] The other indigenous people in the region now began to succumb.

The white settler mode of invasion, violence, and devastation soon fell into a recurrent pattern. Conquistadors had usually not brought wives with them and sometimes married into the local population, increasing the possibility of later creolizations, however limited. Settlers in North America, however, arrived as families, often demanding that their orientations to land and religion receive full hegemony. For instance, the "Scots-Irish," as they were called, were Scottish Calvinists who had been recruited by England to colonize Ireland earlier. Now, centuries later, a famine propelled millions of them to America. They were ruthless settler/invaders, invading new lands in the name of god and the universal good of dividing it up into private parcels of property. They thus defined the collective practices of farming and land management of indigenous peoples to be an insult to civilization itself. First, they would move into a zone that the government had not cleared for "settlement." They would divide indigenous lands, onto which they "squatted," onto numerous parcels of private property. If and when indigenous reprisals occurred they would form private militias and call in federal troops to help them. White militia movements have a long history in the United States.

Ralph Andrist, in *The Long Death: The Last Days of the Plains Indians*, delineates how the pattern set in the east was continued and exacerbated on the plains and far west of the continent. The stories of new settler encroachments after each treaty, met by Indian reprisals, and followed by the entry of the US Army, is repeated over and over. This book is a compelling read, even if its author echoes from time to time Tocqueville's language of forlorn civilizational necessity.[11]

It should be noted, too, how hundreds of American "western" films in the 1950s and 1960s covered these processes. Typically, stout, upright "pioneers" tried to make a new life for themselves on the "frontier." A few white "renegades" might incite Indian "uprisings"; other white heroes would then save both pioneers and Indians from those deceits with the help of misunderstood white women barroom dancers, encouraging the rebels to return to the "reservation" to live peacefully. I attended numerous such films as a youth, moved by the white romances upon which the films focused and vaguely unsettled by the westward march of genocide that set the background to these adventures. The incredible scenes of beautiful landscape setting the background to such scenes doubtless helped to pour a sense of white entitlement into whitened viewers. Such films insinuate unconscious prompts into white cultural life, desperately in need of remedial work. Tocqueville was brilliant in pointing to the cultural power of "mores," sadly callous in his unwillingness to challenge the set he embraced. The mellow Tocqueville, you might say, was an early enabler of later white militia movements. Oppose their tactics, embrace their ends.

Andrew Jackson, of Scots-Irish heritage himself, made his reputation as a soldier who captured escaped slaves in Florida and killed Indians. He broke with the Tocqueville drive to paint white expansion in tones of regretful necessity. When he became president, elected in part because of his fame as an Indian killer, he refused to enforce a Supreme Court decision decreeing that the Cherokee had the right to retain their lands in Georgia and allied areas. Violent militia raids, encouraged by him, overwhelmed the legal decree; Cherokees were compelled to take a deadly march to Oklahoma, where similar pressures would be applied to them later. Since the Cherokee had adopted English land practices and some their religious views, racism now became the key impetus for these land grabs.

As Amerindians retreated westward, temptations of new land, gold, coal, oil, uranium, and space progressively induced first white predatory settlers and then corporations to invade new lands, break treaties, and call in federal forces to protect them. This all occurred under the myth of the American frontier, as the god-given destiny of white America to move ever westward. The myth was supposed to serve as a safety valve for class struggles in the east, allowing each new generation to siphon off discontented workers and poor people to travel to "virgin lands" farther west.[12]

One fateful step in that westward white march of invading settlers, military operations, and eco-damage was the Northwest Ordinance of 1787. Here is how Dunbar-Ortiz describes it:

> Although private property in land had long been a fact of life in Europe, it was demarcated by the contour of streams, rivers, tree lines, rock formations, and mountains and was reserved for the economic and political elite. The United States, being founded as a settler-colonial, fiscal-military state, created something new under the sun, the plat system of privatizing land into marketable units. The Northwest Ordinance spawned . . . a unique surveying method to . . . divide land, transforming it into property for sale and settling. . . . As the US took more land with the Louisiana Purchase, the Oregon Territory, and half of México, the government promised "free land" to Europeans and European Americans for the purpose of recruiting and motivating settlers to squat on Indigenous peoples' lands. With Indigenous resistance to the squatters, the army would be dispatched.[13]

There you have it. The co-original sins of the European invasion: aggression against indigenous people because of their lack of property, agriculture, Christianity, and whiteness; enslavement of Africans for private profit; incredible violence in both instances; and the cultural insistence that land was divinely ordained to be divided into privately owned plots. The violent formation and advance of white settlers meant simultaneously the eradication and enclosure of remaining Amerindians into territorial spaces called reservations. And the "treaties" that created the reservations, too, if and when oil, coal, timber, and uranium interests were at stake, could easily be broken or modified. The pattern spread across the continent.

The End of the Frontier

The frontier myth played a powerful role in the consolidation and supremacy of white, Christian, private propertied, agricultural America. New semblances of it emerge today in white working-class racism and the superrich who seek remote islands to escape a faltering America. Or even the imagination of planetary escapes. But these new, fictive outlets are fantasies; they live on like zombies as more and more people realize in their bones that the capitalist, frontier, white supremacy order not only has been unbelievably destructive but cannot sustain itself in its old terms into the future. Are you that eager to move to California anymore, as it encounters wildfires, energy shortages, drought, atmospheric rivers, flash floods, and water crises? It is no longer a fictive place of escape from class grievances, racial mixing, and other ills. It signifies the death of the violent frontier myth. Many conclude that it is now time to upgrade the six-shooters and rifles of the old west into more potent weapons.

As Greg Grandin asserts in *The End of the Myth*, the day that Donald Trump staked his 2016 campaign around the promise to build a magnificent territorial wall to protect the United States from invading Mexicans and Muslims was the day the old frontier myth, already on life support, expired. The targets remained the same. But the story shifted from white settler demands to expand west by whatever means to demands to protect occupied sovereign territory with a new territorial wall. The agenda now became to stay in place while opposing virulently non-Christian people of color, Blacks, Muslims, efforts to fend off climate change, gender diversity, and attempts to pluralize family and property forms to adjust to a new world. To the extent that old class resentments were once funneled into a frontier myth, the new response expresses an unholy alliance between white neoliberal elites and lower- and many middle-class whites squeezed by closure of old opportunities and resentful of minorities making claims to inclusion in a robust pluralist order.

We now inhabit an era when an aspiration to fascism draws together many old neoliberal elites, insecure working- and lower-class whites, and a looser assemblage of others also disaffected from life for this or that reason.[14] The formula is familiar: you first demand exclusionary rights for your faith, race, fossil fuel, sexuality, property, and/or gender and then insist that you are being persecuted if and when others refuse to grant such an inordinate demand.[15] You forge a political formation in which old professions of ignorance are supplanted by expressions of glee in racism and other depredations of white settler society. The ugly alliance between overweening neoliberals and a white, insecure working, lower-middle class rests in part on the tendency in a stratified society of those on the lower-middle registers of class to focus resentments down rather than upward when things go badly for them, metabolizing preliminary resentments against elites above them into intensification of resentment against those beneath them in a stratified system.

Those who push the cutting edges of white, Christian, capitalist fascism today are ready to enact or tolerate violence against a broadened array of citizens who escape or challenge those molds. The former become increasingly virulent and even prepare for civil war. Implacable pro-gun sentiment, for instance, is driven above all by the dream of fomenting a white revolution, one designed to return to an imagined past.

It is a fragile time. And accelerating climate wreckage—due to a significant degree to white, neoliberal, and evangelical obstinate vetoes of attempts to respond to it over the last few decades—accelerates the fragilities.

To me, the best chance to respond today, among a bad lot, resides in a twofold movement. First, you consolidate a militant pluralist assemblage led by women, Blacks, Hispanics, minority sexualities, dissident Christians, Muslims, and indigenous people to pursue land and wealth reparations, to rework the old unitary property form, to reduce inequality, and to fend off climate change. That means to take on neoliberal capitalism head on. Second, as those agendas are pursued, you also relieve the economic and existential insecurities of the white working and lower-middle class, recruiting a larger section of it to general pluralizing, egalitarianizing policies. Never appease it, but include it in your agenda. You thus seek to break the white/neoliberal/working class assemblage by pulling a portion of the latter out of it. Attempts simply to defeat this constituency have to date backfired big time.

I have found in recent years that diverse minorities eager to join in the first venture are often more reluctant to participate in the second. That is understandable, as an initial impulse. But, to me, the effect is to cut off your nose to spite your face during an era when democratic society is at risk.

A better, if extremely hard to enact, agenda is to rework and pluralize the property form, to redesign the infrastructure of consumption to make it more egalitarian and ecological, to struggle against the multiple forms of racism and patriarchy, to become highly responsive to religious pluralism, and to fend off the growing threat of fascism, all at the same time. I call such an agenda an improbable necessity; it is *improbable* because of multiple institutional demands and constituency obstinacies that work against it, *necessary* because too much of value will be lost unless diverse constituencies pursue such an affirmative agenda militantly together. Young cohorts in each of the potential constituencies are most apt to appreciate and mobilize such necessities.

4
Descartes, Kant, and Amazonian Perspectivism

Descartes, God, and Eminent Causality

It might seem that the next place to turn in an effort to recast western cosmology is the work of John Locke. And that does make sense. After all, his fictive doctrine of a state of nature that is transcended by mixing labor with it, combined with the racism embedded in his support of colonial adventures, does fill the bill. It encourages western readers, for instance, to conclude that Amerindians "lacked" agriculture—that indispensable ingredient in relation to the land—when in fact their practices were merely different and less destructive than the deforestation, division into private parcels of land, plowed fields, and soil depletion so familiar to Europe. Nonetheless, first, others have pursued that trail effectively.[1] And, second, Euro-American orientations to subjectivity, divinity, nature, race and time are intercoded in ways that Kant reveals effectively. Together these practices have helped to justify "tutelage" of pagans and to obfuscate later signs of climate wreckage. In my judgment, Kant, who continues the Augustinian tradition in a new key, presents a powerful reading of each of the above listed themes in relation to the others. And many Euro-American intellectuals continue to be "neo-Kantian" today. Here we will first consider Descartes thought and then explore Kant's continuities and breaks with it.

One link between two historical phenomena—colonialism and the Anthropocene—is that planetary processes marking the Anthropocene often take the form of impersonal, nonhuman circuits that initially distribute their worst effects to the same nontemperate zones that were objects of the second conquest of paganism. The categories of subject, nature, time, class, morality, race, and regularity, to which Euro-centered sciences and humanities have clung for so long, diverted awareness from these developments, even as the climate wreckage rumbled along beneath their feet. Engagements with Descartes and Kant may help us to see how such a complex of assumptions, concepts and existential demands performed those functions, how resistance to changing them became so powerful, and, moreover, how these concept/existential demand clusters set up screens that have deflected official awareness of climate change on the part of the regimes who initiated them until late in the day.

It has often been held that Descartes introduced a radical break into European theo-political thought. After all, while defending an omnipotent God he did attack mercilessly Scholastic ideas of a purposive nature held to reveal darkly God's purposes. But, as Hans Blumenberg has charted in detail in *The Legitimacy of the Modern Age*, the divinely inspired purposive image of nature had already been attacked *within* the church itself by nominalists; they insisted that to confess a purposive world was also to limit the very divine omnipotence also professed. If you discern providential purposes in being to which god is bound you also become a sinner who tacitly denies his power to alter the course of the world at any time. That is, you deny divine omnipotence. This charge, indeed, posed a powerful challenge to Augustinians, Thomas, and others who insisted that god was both omnipotent and tethered to a purposive world.

Blumenberg writes: "The world as the pure performance of reified omnipotence, as a demonstration of the unlimited sovereignty of a will to which no questions can be addressed—this . . . meant that, at least for man, the world no longer possessed an accessible order." And, "Among the propositions of Nicolas of Autrecourt that were condemned in 1346 and that he recanted at the public burning of his writings in Paris a year later can be found the thesis that the precedence of one being over another cannot be demonstrated with evidence."[2] That is, faith now gained even more priority over proof, and to protect faith in omnipotence it became incumbent to break the hold of a darkly discernible purpose and providence in history to humanity that priests could interpret to guide the faithful. An omnipotent god could change the course of discernible history, breaking the apparent flow of a providential course that had been tethered to time itself. The unity of the Catholic world was being challenged: to accept divine omnipotence is to deny a discernible historical purpose in being.

Nicolas of Autrecourt's books were burned by the church in part because they fomented existential uncertainties intolerable to the priesthood and in part because they severely confined its ability to authoritatively interpret the progressive arc of the true faith in history. The affinity between Autrecourt, a nominalist who argued god was not governed by a discernible telos, and Descartes resides in the latter's critique of teleological finalism and his introduction of a hypothetical "demon" who could deceive humans in any way at any time. The difference between them was that Descartes's projection is a fictive demon rather than god itself. Moreover, while Autrecourt propagated an unlimited god to promote greater piety in a highly uncertain world, Descartes deploys his own demon to clear the way for discovery of a slender base of certainty upon which worship of god and mastery of the earth could be promoted together.

Another similarity between them is that while Nicolas was forced to recant in order to avoid execution, Descartes, too, was often on high alert, worried that the church might come down hard on him too. This cautious orientation probably filtered into unconscious restrictions governing his thought as well as his conscious exercise of caution. We readers, too, need to become more alert to how embedded existential lures, anxieties and disciplines prompt and limit conscious explorations, doing so without suggesting that such prompts determine outcomes entirely. Even contemporary theorists in this or that field, for instance, remain subconsciously attuned to how subversive themes they may be tempted to pursue could obstruct their academic advancement or even push them out of the academy—one modern institutional site of much intellectual life.

The institutional matrix in which thought occurs thus becomes insinuated into specific prompts and hedges from which it proceeds. Without such preliminary orientations thought cannot proceed; with them it sometimes becomes less adventurous than the times require. No recent burnings of intellectuals in America to date, but library censorship, book burnings, and attacks on universities are again on the rise.

Descartes, impressed with the recent accomplishments of Galileo in exposing how planets circulate around the sun (rather than the sun around the earth), sought to establish foundations for thought that are certain. His deceiving demon is a ruse of thought designed to carry reflective people through all experiences that can be doubted to one that is indubitable, above doubt. From that base he sought to build a deductive system.

Our sensory experience, he shows, is susceptible to doubt. A stick may appear to bend in the water. A dream may feel real. What is not so susceptible to doubt? Well, that the subject who doubts exists. "I think, therefore I am." What else is certain? That a tight, binary logic is available through which to reach definitive conclusions once that first indubitable experience has been established, and that a benevolent, omnipotent, salvational god can be proven to exist once these two findings are settled.

We will turn to his proof of god in a moment. But first, how does Descartes *know* that a binary logic corresponds to the organization of the world in itself? If you even ask such a question, do you risk falling into a maze of cloudy thinking and unmoored opinions that eliminate the possibility of rigorous thought?

Descartes and many others have said so. But some later western thinkers such as Willard von Quine, Wittgenstein, James, Whitehead, and Nietzsche have contested that very judgment. The shortest route to see how they could do so, perhaps, is to review briefly Quine's version. We will encounter Nietzsche's version later.

The analytic/synthetic dichotomy, which sits at the base of a binary logic, Quine says, does not really hold. Rather than accepting that all statements are divided neatly into those that are analytic—true by definition—and those that are synthetic—testable only through empirical observation—we participate in interwoven "webs of belief" in which some themes are more analytic, some are in between, and others are synthetic or empirical. If a new event disrupts that web—say an untimely earthquake, or a delay in Jesus returning to the world, or the discovery of quantum indeterminacies, or encounters with rapid climate heating—the tendency in both myth and science is to protect higher order beliefs by revising those lower on the current scale of analyticity to fit the new experience.[3] You thus, say, adjust your theology to accept that Christ will return at a later date, rather than allowing the delay to unravel other high order beliefs to which you are existentially attached. What had once been analytic in the Pauline faith—that the second coming will arrive soon—now becomes extended in time to protect beliefs even more dear to devotees—that Christ will arrive and that eternal life is possible. On another front, the rise of apparent quantum indeterminacy may encourage some scientists to posit a multiverse to protect the theme of determinism, saying that every possibility is actuated in some world. So the experience of unexpected events suggests that most of us in fact do follow a more distributive logic in which some elements said to be purely analytic at one time assume a looser relation to faith or science at another. This is an operative logic with shifting degrees of analyticity, rather than one governed by an analytic/synthetic dichotomy.

All this would be unacceptable to Descartes, who joins the indubitable experience that he is a thinking being to faith in the sufficiency of a deductive logic that authorizes "clear and distinct" definitions of things and builds necessary conclusions upon that base.

Let's review now how Descartes proceeds with an ontological proof of god's existence, a proof asserted both to be undeniable and necessary to other philosophical conclusions he advances: mind/body dualism, the authority of binary logic, nature as a deposit of mechanical forces, the disengaged self, and the prospect of human mastery over the earth. He makes it clear that proof of god's existence is needed because some atheists in his regime, and many outside it, doubt the very existence of an omnipotent, omniscient, benevolent god. He may even worry that his own assault on a purposive world in favor of a mechanical image of nonhuman nature and human bodies increases the urgency of such a proof. Since "the argument from design" has now been forsaken, the possibility of secure knowledge that god exists now depends upon such a proof.

How does the ontological proof of god's existence work? The first move is to say that we are able to *conceive* of such a being who is omnipotent, omniscient, benevolent, and salvational. Since I, a lower creature, can at least conceive vaguely of such a higher being, the thought of its existence means that it must be true. But, of course, Anselm had offered an earlier version of such an argument, and one of his critics asked him whether it followed from the fact that he could conceive of an island in the middle of the ocean that the island exists? So, Descartes knows another arrow must be added to his quiver to fix the argument and the conclusion.

The needed arrow is the theme of *eminent causation*. It rests upon an indubitable, undeniable rule. "Now it is obvious, according to the light of nature, that there must be at least as much reality in the total efficient cause as in its effect, for whence can the effect derive its reality, if not from its cause?"[4] Again, "the more perfect, that is to say that which in itself contains more reality, cannot be a consequence of the less perfect."[5] An eminent cause is higher and more complex than the effect it produces or enables; moreover, we must assume that all lower forces and beings can only have their sources in higher causes. The highest cause of all is an omnipotent god who created human beings with souls and lower animals without them. If the assumption of eminent causality is treated as undeniable, the ontological argument for god's existence can now succeed.

But what establishes the certainty and overweening character of eminent causality? Descartes contends that the inconceivability of every other alternative does so. But, clearly, other alternatives have been advanced in the past and Descartes knows it. The idea of emergent causality, in which more complex realities evolve from resonances back and forth between less complex realities until new entities are fomented from below, is one such contestant. Anaximander, Prodicus, Epicurus, maybe Ovid, and Lucretius embody powerful and historically muted strains in pagan thought speculating upon just such possibilities. What Descartes took to be an exclusive assumption was in fact a profoundly contestable one. Other possibilities could be conceived and pursued.

How were such alternative possibilities foreclosed from historical plausibility, then? Here the historical conquest of pagans, the contemporary power of the church, the force of public opinion, and Descartes's own hope to found a new science functioned together to depress emergence as a credible possibility to consider. Negative fears and positive hopes for a new world may have converged in the Cartesian unconscious to mute this option and, perhaps, to discourage its formulation from coming to mind for most of his audience. This was the Cartesian world of preliminary prompts and occlusions. They did not make such thinking impossible, merely less plausible and more dangerous to pursue.

Giordano Bruno was burned at the stake for advancing views about emergent causality. That makes people think, or, better put, stop thinking too far outside the orthodoxy of the day. Lucilio Vanini, to take another instance, was born in 1585, and soon published dialogues against atheism and skepticism. But the official views he supported were, to close readers, effectively undermined by the power of the arguments by characters in his dialogues he officially demeaned. The church caught on. "Vanini was handed over to a court and found guilty of atheism, immorality against nature, and seducing the young. He was sentenced to be taken to the Place du Salin in Toulouse, where his tongue was to be ripped out, after which he was to be strangled and burned. The sentence was carried out on February 9, 1618. Witnesses of the execution said that they were unable to forget the dying man's agonized screams."[6] That concentrates the mind; rather, it primes it in some directions over others.

I am not suggesting that Descartes was a closet atheist or even a deist, though those suggestions have been made by others. Why not? He needed to prove an omnipotent god to establish other conclusions so important to him. I am suggesting, however, that the conjunction of disparate historical forces—most notably church executions of heretics joined to his own quest to support a new deductive science of nature—discouraged active exploration of possibilities that would, if they gained credibility, call the ontological necessity of eminent causality into question. The idea of emergent causality, rather common in recent thought in quantum physics, evolutionary biology, planetary processes, and societal changes, was at that time historically depressed rather than logically eliminated. The first conquest of the pagans in Europe, indeed, helped to support its silencing.

Few, then, would publicly impugn Descartes's rejection of that assumption in his day on that terrain. They, too, could run into trouble, as Spinoza himself did during the same era for skating too close to heresy. And when Spinoza died, the church commissioned a cadre to try to intercept the manuscript of the *Ethics* before it could reach a publisher.

We are thinking/feeling beings with preconscious prompts to thought insinuated into us as we grow up; we must undergo critical work on such projections in order to think differently—as Descartes in fact did in his herculean effort to break with a purposive view of the world which had been drummed into everyday experience. Such affectively imbued prompts are not merely individual. They are culturally instilled through ritual experience, through absorbing priorities of dominant institutions into daily routines, through formal disciplines, and through everyday transmission by way of looks, intonations, gait, caresses, timely hits, and frowns. This is the visceral register of cultural life to which anthropologists have been alert in studying nonwestern cultures; it has, at least, until recently, been given less prominence in self-conscious reflections about the west itself. We today may confront preconscious prompts to sociocentrism, for instance, that unconsciously bound diverse thinkers such as, say, Keynes and Hayek together along one strain of thought even as the latter fought militantly against the former along other strains. Sociocentrism formed a common pivot of modern debates which, for that reason, was difficult to subject to comparative scrutiny. A cultural unconscious infusing diverse thinkers and institutions encouraged it. That unconscious is not to be challenged merely by becoming more self-conscious. It also requires, as Nietzsche, Foucault, and Jane Bennett too have seen, experiments with counter-*techniques* to disrupt old, preconscious prompts and to enable new ones.[7]

The deductive Cartesian system, then, was imbued with and encouraged by unconscious strains of prejudgment and plausibility. Prior scholastic doctrines, strains of which Descartes now opposed, shared the theme of a binary logic with him. And species evolution was radically implausible to Descartes, to the point of appearing to be a logical absurdity.

With these preliminaries behind him, Descartes could now support other conclusions for which he has become famous: mind/body dualism, nature as a set of mechanisms rather than living forces that strive, the human body as a mechanism, and a future science built around mastery by human minds over an inert earth.

His mind/body dualism has received a lot of critical attention, particularly by new turns in neuroscience.[8] According to it, mind is a disembodied soul and body is an extended mechanism subject to scientific analysis. The soul is the recipient of eminent causality. But the body, while it relays humors to the soul, is not itself replete with clear and distinct ideas worthy of close philosophical attention. Descartes knows he has trouble showing how these two disparate "substances," which must interact, in fact can do so. The dialogues with Princess Elizabeth provide a fascinating series of attempts to show how two separate substances—one marked by thought without extension and the other by extension without thought—can interact.

There is another aspect of Descartes's philosophy that must be placed on the table too, partly because it was bold to advance it at the time and partly because it contrasts with orientations to be considered later here. His philosophy of time.

God is not the vehicle of an inherently purposive time discernible to us. Nor does he participate in a time that exceeds him, as many pagans had thought. Either theme would undermine divine omnipotence. Rather, *time as a series of distinct instants of succession itself depends upon the willingness of an omnipotent god at each instant to will the next instant.* Life, under this god in charge of time, is thus fragile. Without him it stops: "for the whole duration of my life can be divided into an infinite number of parts, no one of which is in any way dependent upon the others, so it does not follow from the fact that I have existed a short while before that I should exist now, unless at this moment some cause produces and creates me, as it were, anew or, more properly, conserves me."[9]

Time is thus eminently caused by a god who presides over existence, singularly deciding to stitch instants together . . . or to bring them to a stop. Here, we feel more dramatically why god is so critical to the Cartesian system. We also sense how fragile the world feels within it, dependent upon the unfathomable will of god to decide whether to continue or stop time at any "point in time." Descartes's philosophy of time constitutes a break with scholasticism; it has been carried forward, in different ways, in the thought of Newton and Einstein. To me, the theme draws together Descartes's sense of how fragile existence is, how crucial god is to humanity, and how important it is for science to enunciate god-given natural regularities to improve the quality of human life on earth. It is not that these three imperatives determine one another—more that the last one becomes more urgent existentially to pursue once you are committed to the first two.

Animals, to Descartes, as parts of nature without souls, are mechanisms without high cognitive capacities. If they possessed higher capacities, they would be able to speak, and he has not located any that do. They are thus "thoughtless brutes," susceptible to human use. Descartes would thus foreclose ecological studies that explore the complexities of whale, fox, octopus, fungal, and bacterial life, let alone those viral species crossings that are so pertinent again today. Let alone, too, those gut/brain relays through which mood-dependent thinking takes place in human beings, that is, through which internal, nonhuman agencies and forces filter into its conscious agency. He also establishes a natural hierarchy of human beings in which those with great logical skill are most intelligent, others are only equipped to navigate everyday routines, and others yet are stuck in fantasy and dream worlds that make them need higher-order governance.

Little hint of democracy in Descartes, then. Does he apply such a formal hierarchy of human beings to non-Christian cultures? I have not located places where he does so, but the implication hangs over his philosophy like an upcoming storm over the dawn. He is pressed by his system to construe regimes oriented by multiple gods to be lower in sophistication than those that profess a single, omnipotent god; he can't help treating a cultural regime anchored in belief in an austere, distant, omnipotent being as superior to those governed by a superstitious view of the vague discernibility of divine purposes in the march of time itself. He is thus also pressed to construe minor pagan philosophies of emergence to be inferior to recognition of the necessity of eminent causation. For within his system there is no divine guarantee that temporal trends that have persisted will continue to do so into the future. God itself does that—or not—at every instant.

Finally, let us note how Descartes, after *disengaging* human beings from nature—through their uniqueness on earth as beings with souls and (in some cases) high logical capacities—then elects an elite to develop a deductive science to master and dominate the nature from which we are disengaged. The defining objective is made clear in *The Discourse on Method:* "by knowing the nature and behavior of fire, water, air, stars, the heavens, and all the other bodies which surround us, as well as we now understand the different skills of our workers, we can employ these entities for all the purposes for which they are suited, and so make ourselves masters and possessors of nature."[10]

"Knowing . . . nature," "understand the skills of our workers," "employ these entities," "make ourselves masters . . . of nature." A disengaged, aristocratic culture can develop instrumental dispositions and knowledges to conquer nonhuman nature and to regulate the mechanistic bodies of human workers. By grasping the mechanical nature of both such entities better, by perceiving how the bodies of workers can be deployed to increase productive power, "we" can construct a master technology of bodies and nature that increasingly serves human welfare through control of nature. As long, of course, as a gracious god continues to renew those divine links between one instant and the next, second by second, day by day, year by year, decade by decade.

Recall how Columbus was a figure poised between the finalism of the European past and an instrumental orientation to nature. The latter found expression in his insistence upon taking possession of new territories and exploiting the nonhuman and (to him) barely human resources there. Descartes solidifies the latter side of the Columbus orientation as he denies the first. He is thus a figure who forges links between those orientations that developed during European colonization of nature and non-Christian peoples and those by which Euro-American capitalism has gone on incredible sprees of production, consumption, exploitation of workers, extraction of fossil fuels, and exploitation of nature precipitating the ravages of the Anthropocene.

There was little in Descartes's mechanistic view of nature that led him to anticipate the future volatilities capitalist projects of domination would one day unleash. Deaf with respect to the resistance of colonized peoples, blind to the way practices of mastery could unleash the volatility of large nonhuman forces, and tone-deaf to a multitude of micro- and macro-thinking life forms within and outside human life, the Cartesian legacy plants the seeds of planetary devastation and its own unraveling. Grounded simultaneously in existential anxiety about a distant god's control of time and in hubris about the human relation to nonhuman nature, the Cartesian system links two cultural mentalities and two Euro-centered periods: the period of direct invasion and colonization and a period that instigates a new phase of Euro-American imperialism over nontemperate zones through impersonal planetary circuits. More about the second movement later.

Kant, Descartes, Augustine, and Prodicus

Kant is a system builder who sets severe limits around the boundaries of that system. He seeks to show how each part of it fits inexorably with the others. So he gives considerable priority to knowing and demonstrating in his image of thinking, whereas others could also emphasize within thinking itself the value of encouraging unconscious thoughts to surface, speculation, promoting real creativity, and even inspiring others to explore new avenues. As expected for the author of a philosophical system, Kant's thought is extremely complex, even sometimes convoluted in its attempts to fit everything into its proper place. To hold the three major offices of reason together (understanding, practical reason, and teleological reason) while incorporating findings from Galileo and Newton about nature into the system, he is recurrently called upon to make new distinctions and, soon enough, to refine those too.

It is exhilarating for those who admire systems—and who think that arguments designed to prove universals express the highest form of thought—to study Kant. It is frustrating to do so, too. Hence some might discern in Kant, Kantians, and neo-Kantians a certain existential anxiety, expressed, first, through repeated attempts to hold a rickety edifice together and, second, through intense attacks on those held either to misunderstand the system or, more belligerently, to rebel "irrationally" against it. Kant loves reason, as he comprehends it.

Those who read Kant to think and rethink characteristic western debates about the relation of the cosmos to culture and politics might get initial bearings by comparing him to three figures we have previously engaged: Descartes, Augustine, and Prodicus.

Descartes and Kant share at least two fundamental affinities across their differences. They both seek, first, to square some version of Christian doctrine with a conception of nonhuman nature reshaped, first by Galileo (in the case of Descartes) and later by both Galileo and Newton (in the case of Kant). To do so, each develops an image of nature that is subject to something close to mechanistic laws, laws set in contrast to the human mind or soul that is itself capable of freedom, of moral obedience, and of being lifted into eternal salvation upon death. But each finds a different source for the needed assurances: Descartes grounds the assurances *objectively* in clear and distinct ideas; Kant grounds them *subjectively* in a set of necessary *postulates* we cannot avoid making if we are to be consistent, rational beings. Both Descartes and Kant assume that the European societies of their day have attained a higher level of civilizational achievement than those at other times and/or in other zones of the world at the same time. This second affinity is something argued more explicitly by Kant, whereas it hangs over Descartes's thinking like a clear moon shining over one area of the earth more brightly than others.

Kant, then, saves Cartesianism largely by translating its simple tie to objectivity into a series of necessary elements of human subjectivity. Hence, he must *postulate* a thing in itself but *know* nothing about it "in itself."

Augustine and Kant may seem rather far apart when you compare them only within the Christian tradition broadly considered, and if you take at face value Kant's reading of the matter. Augustine's faith is grounded through confession of the sanctity of Christian scripture (revelation) and, if and when you are chosen, poured through divine grace into the interior of the soul. Augustine confesses faith in free will and reads Genesis to explain how that pure will only became divided against itself after the fall of Adam and Eve.

Both Kant's god and his orientation to free will are asserted to be necessary, subjective postulates you must accept only *after* grasping the apodictic character of morality to take the form of laws you are obligated to obey. These former derivations, for him, thus take precedence over both revelation and confession. So Augustinian theologians are often critical of Kant, and Kantians often dismiss the theological directness of Augustine.

Augustine's image of nature, too, is different from the image Kant adopts—though the distance may contract upon Kant's late study of *The Critique of Judgment*. To Kant, nature in the first instance must be postulated to consist of a set of laws susceptible to scientific understanding, even though we must also postulate those laws to be undergirded by a divine intelligence that grasps "the thing in itself" perfectly. Augustine is more direct. He confesses an intrinsic purpose into nature and also confesses that large aspects of nature are divinely endowed with human dominion over them, exempting only the earth in its entirety, the sun and stars, and (by implication) climate from human dominion. Moreover, while Augustine confesses an original sin that profoundly compromises free will after the fall, and whereas he holds unfathomable, divine grace to be essential to salvation, Kant denies original sin. Eventually, though, Kant does talk about how the human will can become profoundly divided against itself in this tangled life, even though it always remains beyond mechanical determination. Kant is thus finally moved to reinstate—in his own way—a "hope" for divine grace in corrupted human beings, a hope for grace that can aid the will in pulling itself out of a corruption, an act it is often not able to do by itself. Kant and Augustine thus move close in their conceptions of the will and the priority of humans over other beings of the earth. Freedom, grace, and dominion, though in different ways, are important to both thinkers.

Judgments of how close Kant and Augustine are depend on the standard against which you compare them. *Within* the Christian tradition they appear rather different, one more direct in his confession of faith, the other tied to postulates that are subjectively necessary but not known to be true in themselves. When you compare both to pagan philosophies of immanence, emergence, evolution, and an ethic of cultivation, however, these two figures move closer together. It is fortunate that a compelling book by Gordon Michalson, titled *Fallen Freedom*, helps to sort out these complexities and to expose affinities many had previously underplayed.

What about Prodicus, then? From overt things Kant has said about nonplatonic, pagan thinkers such as Epicurus and Lucretius, we can see that the Kant of the critical period—that is, Kant after rising above his early thought—would find the thought of Prodicus to be unsophisticated and unsystematic. The latter, to him, practices a mode of prephilosophy. Since Prodicus (we can surmise) neither postulates a god, nor derives morality from a matrix of universal laws, nor works to prove a system through an interlocked series of transcendental arguments, Kant would define him as a deficient thinker. He would also strive to catch Prodicus in a series of putative contradictions.

Most radically, Prodicus seems to have been a philosopher of *emergence* who speculates that life itself emerged from protean modes of nonlife and after that evolved in various directions. There is thus way too much contingency, emergence, and evolution in the world of Prodicus for Kant. The judgment between them depends to a considerable degree on whether you conclude that Kant's transcendental arguments work and, within that matrix, whether you take Kantian images of morality and time to be firmly established. It is clear, too, that from the vantage point of Prodicus, Kant and Augustine embody remarkable affinities across their differences. They both project an omnipotent god; each either assumes or postulates the primacy of eminent causality; they both oppose the idea of species evolution; they both ground morality in a higher command; and they both set a perfected Christianity at the apex of the good society.

To elaborate one line of difference, it might be useful to compare Kant and Prodicus on earthquakes. Such eruptions were common parts of experience in the world of Prodicus, less so in that of Kant. Neither Prodicus nor Kant, recall, had access to a theory of tectonic plates, a theory that only achieved fruition in the 1960s or so! We can suppose that Prodicus located many of his findings about a bumpy world in the domains of experience and speculation.

When the devastating Lisbon quake exploded in 1755 on All Saints Day, Kant was pushed to action. He denied, against most Catholics and Protestants of the time, that it had anything to do with divine punishment. He sought natural causes for its occurrence, constructing a hypothetical theory of caverns and chemical reactions below ground that could cause these eruptions in accord with a model of lawlike natural processes.[11] The Lisbon shock, indeed, may have been one of the things that motivated Kant to move his earlier, more direct theo-philosophy of a teleological world governed by providential purpose to a more indirect, critical philosophy in which, to sustain an image of ourselves as free, moral agents, we *must* adopt a series of postulates about ourselves and the world that underpin our capacities to study the laws of nature and to act morally. We will pursue those postulates soon.

But it is pertinent to see that Kant would not embrace a Prodicus image of the world as periodically rocky in itself. Or the emergence of life from nonlife. Or an ethic of cultivation rather than one of universal law. If compared to Jocasta or Thucydides, Kant would be even more adamant in postulating a world that is providential in the last instance over one that is more capricious and volatile in its relations to humanity.

To Prodicus, an omnipotent god is neither an internal experience securely grounded in confession and revelation (Augustine) nor a necessary postulate recognized by mature moral subjects (Kant). Given the protean character of the materials from which the earth emerged, Prodicus seems to speculate, humans and other species may well have emerged without divine guidance; civilizations and ethics could evolve further without it too. So Prodicus gives far more range to *theoretical speculation* than Kant does. He would doubt the putative power of Kant's transcendental arguments, to be discussed soon.

The ethical life, to Prodicus (we might surmise), is neither the upshot of a necessary command nor reducible to mere preference—the latter being the alternative Kant seeks to pin on many of his philosophical opponents. Ethics, rather, emerges from a give and take between people who internalize a degree of care for one another and the earth at their mother's milk, if they are lucky. If that internalization is absent, well, then, a stage is set for tragic results. To Kant, though, any notion of ethics—or morality as he calls it—is anathema unless and until it is grounded in obedience to universal law. To him, to ponder tragic possibility is to remain stuck in images of gods, subjects, and time historically superseded. He, again, rejected evolution, though within the human species he did (in his early thought) postulate a process of racial and civilizational "devolution" from a single, superior strain of white humanity to a racial hierarchy he observed around him. Kant's images of the subject, morality, time, civilizational hierarchy, and natural philosophy, as we shall now see more closely, are bound together.

Kantian Postulates of Freedom, Nature, Markets, and Time

We will not explore how Kant establishes the intuition of time as succession in the First Critique.[12] His judgments of other civilizations and nonhuman processes gain their most purchase at a later moment. So, let's start with the lynchpin of the positive Kantian system and proceed from there. How do we *know*, he asks, that morality takes the form of universal law? We don't (and can't) know it deductively or by empirical or historical investigation, says Kant; rather, we *recognize* apodictically that it does so. Once that recognition is secure, a whole set of necessary postulates and postulate-like hopes and projections follow. They do so through a series of interlocked transcendental arguments. A transcendental argument proceeds roughly like this: You first identify an experience that is undeniable and indubitable. You then ascertain what presuppositions must be accepted to sustain that indubitable experience. And you finally accept the latter finding as a necessary postulate to embrace. You don't know it in itself or directly to be true in the first instance; you postulate it subjectively in the second instance to be necessary to us.

So let's listen to a formulation of apodictic recognition in the sphere of morality. Here is what Kant says:

> For whatever needs to be drawn from the evidence of its reality from experience must depend on the grounds of its possibility on principles of experience; by its very notion, however, pure yet practical reason cannot be held to be dependent in this way. Moreover, the moral law is given as an apodictically certain fact, as it were, of pure reason. . . . Thus the objective reality of the moral law can be proved through no deduction, through no exertion of the theoretical, speculative, or empirically supported reason. . . . Nevertheless, it is firmly established of itself.[13]

We don't learn from empirical observation that morality takes the form of universal laws we are obligated to obey, nor do we deduce it from other facts, nor do we merely accept it as a reasonable set of conventions, nor do we ground it directly in scriptural faith. No. First, we recognize apodictically that morality takes the form of law and then deduce other implications—postulates—from that recognition.[14]

But what, you might ask, about non-Christian cultures that do not harvest the preliminary cultural seeds for that apodictic recognition, even what about those individuals and minorities within Christian societies who contest it? We will return to this delicate issue soon. But for now we can say that, to Kant, non-European cultures fall below a high level of civilization if or when they do not attain such recognition; their capacities are clouded by myths. And wayward individuals within Christendom who fail to do so express a corrupted will. He is very severe on those convicted of this latter failing. Also, what if some of these other cultures convey cosmologies that treat the planet to go through periods of radical tumult? They are problematical too on this account, as we shall see.

Now apodictic recognition sets the base upon which a series of postulates with different degrees of necessity are established. A series of transcendental arguments thus now become formulable, because the first indubitable recognition has been attained. Let's consider several of them.

Postulates of god, the free moral subject, and eternal salvation. These first three postulates are so close to the apodictic recognition that they can almost be said to be part of it. To recognize morality as universal law is to postulate a divine being who is the author of such laws. Key here is that the necessary postulate of god does not precede recognition of morality; it, rather, follows from it. Such a judgment is tantamount to giving philosophy priority over theology. And it got Kant into considerable trouble in his day with both ecclesiastical authorities and the Prussian, Christian state. Christian theology and ecclesiastic practices, says Kant, *prepare* people to sharpen apodictic recognition, one that gives priority to morality over "religion," as he called it. But these religious traditions and practices cannot secure their own necessity, except through philosophy.

Closely connected to the first implication is the postulate of free will. The recognition of morality defines us as agents free to identify the moral law and free to obey or disobey it once various tests have been run to determine whether a potential law is indeed obligatory.

The postulate of progressive time. If you study history empirically, you may observe beneficent things at one time, indifferent things at another, and terrible things at yet another. That Lisbon quake, for instance, seemed to constitute a break in a projected flow of progressive history. But even though such events are not to be denied themselves as facts, it is also morally necessary to *postulate a general order of time that is progressive.* For if the moral law is to be intact as law, it must be realizable in principle through time. To meet that condition you must anticipate the progressive realization of morality in history. Otherwise you might recognize morality in one gesture and deny its historic realizability in the other. You might become a cynic.

So it is necessary to postulate history as progressive and set on a single trail, heading slowly but inexorably toward a future implicitly lodged in the composition of the moral laws themselves. This is a radical move by the meticulous Kant, one which like the first two, will require later elaboration and clarification.

But it must initially be discerned in its purity. Here is one formulation (the word nature here is not nature as studied empirically but the nature postulated to be filled with providential purpose):

> I will thus permit myself to assume that since the human race's natural end is to make steady cultural progress, its moral end is to be conceived as progressing toward the better. And this progress may well be occasionally *interrupted*, but it will never be *broken off.* It is not necessary for me to prove this assumption. . . . For I rest my case on my innate duty . . . so to affect posterity that it will become continually better (something that must be assumed to be possible).[15]

To secure the apodictic recognition of morality we must now, then, postulate time to head toward a human horizon of progressive moralization whose content we can now discern only imperfectly. For humans are the only earthly agents Kant recognized. To refuse that postulate would be to start to unravel authoritative morality from the outside in. Why? Because *ought* implies *can*, to lose the *can* is to jeopardize its obligatory form. To protect the ought/can relation it is thus necessary to postulate progressive, civilizational time. It is your "innate duty" to project such a postulate.[16]

Kant soon concludes that this lived postulate receives subsidiary support from empirical experience, from little "signs" that connect actual historical experience to the postulate sufficiently to protect the postulate itself from a tragic unraveling by events. For Kant resists with all his might—and through the entire edifice of his philosophy—projection of a world replete with tragic possibilities. Tragedies that do occur are due to moral failures. Tragic wisdom is thus something pursued at a lower phase of historical progress, as Kant makes clear in his critique of a contemporary tragedian.

The postulate of impersonal tendencies within both markets and nature to progress toward a more perfect moral commonwealth. Every individual, and every republic, is obligated to pursue a world that becomes progressively moral. This is a necessary "as if" to pursue, personally and civilizationally. But many fail to do the former through weakness of will, and even more important, *the intentional reach of individuals and entire regimes in this respect falls well below the historic need.* So now we must project another "as if" postulate. We must act as if *impersonal forces in both nature and culture function to make up the historic deficits of human capacity and reach.*

> But now nature comes to the aid of that revered but practically impotent general will. . . . [In a world of conflicting national drives,] one inclination is able to check or cancel the destructive tendencies of the others. The result for reason is the same as if neither set of opposing inclinations existed, and so man, even though he is not morally good, is forced to be a good citizen."[17]

Here Kant moves close to Adam Smith, who posits impersonal market tendencies to attain rational economic outcomes that no central power or individual entrepreneur intends or could realize. But Kant's version, first, takes the form of postulates rather than assertions of empirical tendency and, second, is more demanding. That is, this postulate is not only projected into culture, it is projected into nature too. So,

> if we reflect on nature's purposiveness in the flow of world events, and regard it to be the underlying wisdom of a higher cause that directs the human race toward its objective goal and predetermines the world's course, we call it *providence*. We cannot actually have *cognitive* knowledge of these intricate designs in nature . . . but we can and must *attribute* them to objects only in thought so as to conceive of their possibility on an analogy with mankind's productive activities. . . . From the practical point of view . . . it is represented as a dogmatic idea and it is here that its reality is properly established.[18]

The "as if," therefore, resides in necessary postulates that apply to nature, culture, and time alike. The meticulous Kant does not leave a stone unturned in his efforts to support an apodictic image of morality. Along the way he may disclose the fragility and insecurity of the image of time he projects.

Postulates of civilizational and racial hierarchy. Kant postulates different levels of attainment of the universal among disparate civilizations and races (as he says) at any given time. If the universal goal is attainment of a purified

Christianity in which all recognize and freely obey the moral law, in Kant's day, and during other times too, disparate regimes will be arrayed along divergent levels of progress toward that very end. So Kant postulates different civilizational and racial levels of attainment.[19]

Non-Christian, pagan civilizations and cosmologies are lower on the historical scale of achievement than Christian commonwealths, so some races need more "tutelage" than others to approach the universal ideal. A universal Caucasian race is projected by the early Kant to have degenerated in warmer climates, so that now multiple races form a hierarchy among Caucasian, "Negroid," Indian, and Asian peoples. The progress of lower races to a higher level means that they must be tutored to rise to it, reminding you perhaps of Augustine and the early Las Casas on this score.

These ranking judgments, however, are not mere prejudices of Kant that could easily be lifted while retaining the rest of the system. The rankings make sense of the gap between the projected moral universal *and its variable degrees of attainment by different peoples at each historical juncture.* The recognition that morality takes the form of universal law, for instance, is most apt to take hold in Christian cultures, even though the "ecclesiastical" version of Christianity itself needs to be lifted to a higher level in western life. Indeed, Protestant forms of Christianity are closer to the ideal than Catholic modes.

At this point, at least two things stand out. The Kantian system is magnificent from one perspective, in that its brilliant author is alert to what is needed to sustain its delicate architecture. The result shows readers how a systematic philosophy can be constructed and protected. But those achievements also form part of the problem. They carry a racism of colonial tutelage with them; and they are also ill devised to open critical reflection into the progressive, linear, and singular image of time invested in capitalist, church, and political institutions in the west. They rather re-enforce them. Thus they are thus not well devised to encourage future generations to discern the rise of climate heating. If, for instance, you postulate the gradual advance of civilization, rather than, say, contend that different timescapes (a term to be discussed later) may turn in different directions under new and unexpected pressures, you are less apt to ask whether the market-like "advances" experienced through the march of extractive capitalism carry destructive seeds of the Anthropocene with them. Kant projected too much faith into the rationality of markets. Later followers focused on that assumption are less apt to *detect* early signs of climate wreckage within the postulated advances of nature and time. The postulated trend is, after all, one of impersonal, gradual progress supported by nature and god.

So, a system composed of universal morality and necessary postulates of singular time, nature, race, and civilization carries unfortunate tendencies—

though there remain subordinate elements in Kant's thought that deserve respect. Think, for instance, of how he studied the empirical sources of that Lisbon quake when so many other contemporaries construed it to be a punishment by god. Think, too, of how his critique of ecclesiastical Christianity opened a door for other Christians to treat the Christian creed itself as a contestable faith, one worthy of participation in a pluralistic society but not of setting the universal matrix of public culture. Kant opened doors he could not walk through without cracking his system.

Think, too, of how Kant always insisted that negative critique is not enough. It must be joined to positive formulations, even if they are difficult to offer. For mere critique leaves people at an impasse about what to do; moreover, prompts and premonitions from old cultural perspectives you are criticizing are apt to remain embedded in the soft tissues of life unless a positive alternative is pursued. Experimental behavior changes prompts.

How, though, could the Kantian problematic become pluralized, rather than merely either simply rejected or subjected to critique? The problem with simple rejection, again, is that some elements you now consciously reject can slip back in as remnants unless they are subjected to both close critique and alternative positivities. And, as we shall now see—subjected to gymnastics that may help to replace some old remnants with others.

Postulates, Critique, and Gymnastics

Several Kantian postulates need to be challenged by positive alternatives. The idea is not to eliminate his creed, but to replace it as an undeniable subjective, civilizational system. To acknowledge more profoundly the contestability of those themes while acknowledging that many may reasonably have faith in them.The themes in question are the idea that morality must take the form of law, human exceptionalism, the primacy of subject/object duality, the insistent postulate that time follows a linear path of progress, the postulate of progress supported by impersonal nature, and a racial/religious/civilizational hierarchy sustained by the entire complex. Let's start with the first, the "lynchpin" that helps to tie the others together into a set of existential imperatives.

"We" recognize, says Kant, that morality takes the form of law. But some peoples, even some minorities within his own culture, do not express such recognition. Some European pagans, for instance, had grounded morality in a mode of cultivation, whereby the self works upon some strains in its prior subjectivity to render it more presumptively generous in relations with others, and a civilization does the same. Epicurus provided a fine example of that approach. One modern example of such an orientation can be found in Michel Foucault, and other Europeans such as Diderot and Hume pursued such an agenda in their times and places.[20]

Even Kant finds a minor but critical place for cultivation or, as he called it, "gymnastics." Young people in whom the juridical recognition of the law-like character of morality is initially weak, he says, must be inducted into sharpened recognition through gymnastics, that is, inducted into educational schemes, modes of repetition, light reprimands, carefully stated compliments, and so on by their teachers that help to install the desired recognition more securely into the presumptions of being. "Certainly it cannot be denied that in order to bring either an as yet uneducated or a degraded mind into the path of the morally good, some preparatory guidance is needed to attract it by a view of its own advantage or to frighten it by fear of harm. As soon as . . . these leading strings have had some effect, the pure moral motive must be brought to mind."[21] In other words, apodictic recognition must be sharpened in the young and wayward through cultivation and arts of the self. Moreover, collective tactics are needed to lift entire civilizations lower on the scale of progress to a higher point of recognition. Cultural tutelage of lower civilizations by higher civilizations.

Kant's official view is that such gymnastics work on the *sensuous body*, softening it up to make it more receptive to the moral law once recognition of it as law has been sharpened. A brilliant move in his philosophy, reminiscent in some ways of Augustine's behavioral recipes to draw monks and sisters more fully into the faith, or even of Nietzsche's "arts of the self."

However, gymnastics, once their pertinence is so vividly acknowledged, can also be read another way. Kant himself, for instance, was raised as a Pietist, a branch of Lutheranism that emphasized authoritative, disciplinary tactics designed to pull young people through crisis to more radical conversion that installed faith in Christ more deeply into the self and ethos of the community. Devout modes of collective discipline, augmented by lifelong modes of self-monitoring and discipline. If Kantian postulates are successfully reinterpreted to be contestable projections, the specific gymnastics he embraces can now be treated to be culturally embraced disciplines that, over time, become operational habits of orientation through repetition. The specific prompts into which he was inducted could have been otherwise.

So we now offer a competing reading of these devout disciplines and tactics—or gymnastics. Gymnastics are indispensable to culture in some form. But rather than opening the devout to apodictic recognition of that which already slumbers there *implicitly*, they can also be interpreted as repetitive disciplinary modes that sharpen a set of contestable cultural premonitions already encoded imperfectly in bodily prompts and cultural routines of a specific civilization. And then lifted into more refined practices and theories. Rather than being guided by necessary postulates illuminated by his philosophy, Kantian gymnastics, then, can be read to both imbue prompts to what Kant calls apodictic recognition deeply into selves and cultural routines (prayer, devotion to the cross, deference to the pastor, confession of apodictic recognition, and so on). Such preliminary prompts, once installed, then prepare the way for more reflective philosophical elaboration and acceptance of them, that is, of Kantian philosophy. Such a nexus shows the Königsbergian and the confessional disciplines of the bishop of Hippo to be closer than they might appear at first glance.

Don't forget the indispensability of gymnastics when thinking about Kantian philosophy. They can be both appreciated and treated otherwise: for example, to educate pluripotential possibilities of affect-imbued thinking into alternative channels of prejudgment and existential hope. That is, the specific role and implications Kant attributes to them turn out to be speculative and contestable, even while *some* such preliminary practices and prompts always do subsist in cultural life.

Kantian recognition and postulates, on this reading, can now be defensibly read to be his specific philosophical refinement of a set of cultural prompts with which he was imbued as a young Christian pietist. The prompts carry a degree of efficacy, providing preliminary suggestions which bolster the plausibiliy of the philosophy, though not complete determinants of it. Such a conclusion does not erase Kantianism as a contestable creed; it, rather, smudges the apodictic starting point from which its *necessity* and *universality* were projected by him and others. For, remember, not everything is clear and distinct in itself. Some processes are cloudy and murky in themselves, such as clouds, fog, mist, swamps, viral crossings, and fuzzy ideas on the way to consolidation. Pluripotential preorientations to thinking and judgment very often assume cloudy forms at the outset.

Kantianism now becomes a contestable theo-philosophy linked both to a set of arguments and to specific cultural disciplines and gymnastics that have helped to prompt it. Now the need to dig more deeply and comparatively into pagan, non-Kantian traditions within and outside Europe becomes more apparent, doing so to explore more carefully whether some modes of Euro-American disciplinary inductions ("gymnastics") need to be modified in the light of past horrors they helped to authorize and the unexpected future looming before these very cultures. For example, the Kantian postulate of a cosmos and planet predisposed to us in the long run can now be thrown up for grabs, opening the door to explore insights, say, from an Aztec cosmology explored in the previous chapter or of time as a bumpy multiplicity to be pursued in the next. Both in the light of new, unexpected experiences and renewed engagements with a muted past.

Indeed, Hesiod, Sophocles, Prodicus, Ovid, and Aztec cosmologies can now be received as competing ways—with variable degrees of overlap between them—to launch such explorations. Not necessarily to complete them as found, but to launch them. As we proceed down this latter path, it is wise to recall again how and why the wise Augustine was so committed to such a detailed series of inductions and strictures with respect to monks, nuns, heretics, women, and pagans. These inductions and silencings helped to pave the way for cultural universalization of his confessional god. We will pursue Amazonian perspectivism to close this chapter and time as multiplicity in the next. Are hegemonic Euro-Americans now released to learn more from such lived alternatives?

Amazonian Perspectivism and Nonhuman Subjectivities

One attempt simultaneously to challenge Kantian apodictic recognition, to pluralize the Kantian image of moral subjects, and to project subjectivity well beyond humans alone has been launched by Eduardo Viveiros de Castro. He does so through a sympathetic examination of Amazonian cosmologies. As I read him, he remains alert to key Kantian imperatives, contrasting and contesting them through reference to the lived cosmologies of Amazonian cultures. At least I will mark the Kantian alternatives as we proceed.

One advantage of such an endeavor is that it allows and encourages us to consider alternatives to several Kantian imperatives together. For Kantianism is an intricate system; a critique of this or that theme within it can readily be overwhelmed by reference to how the critique "contradicts" other assumptions held to remain intact. Many westerners have been imbued with the very prompts Kant celebrated and enacted. So we may tacitly adopt this or that assumption in the Kantian complex as we strive to modify another. Kantians and neo-Kantians love to hurl the charge of self-contradiction against critics. It aspires to be a system, after all.

For example, if you challenge the universal, law-like model of morality, a Kantian or neo-Kantian will surely assert that you have just reduced yourself to treating morality as an expression of "mere preference," the favored alternative Kant himself identified and criticized when encountering opposition to his view. I have confronted that charge myself a few times over the years. Kant indeed himself attributed such a stereotyped alternative to Epicurus and Spinoza, though it is highly doubtful that either in fact reduced morality to mere preference. They both moved closer to what I have been calling an ethic of cultivation than to one of ethics as the enactment of mere preference. Similarly, if you challenge the Kantian projection of linear, progressive time without exploring other aspects of his doctrine, it will be said that you have become a nihilist. For you are attacking explicitly your own implicit image of time, one that cannot be denied without self-contradiction. The elements in a worldview, as the brilliant Kant knew so well, form an intercoded assemblage or cluster.

So the Viveiros de Castro task now becomes to challenge a large cluster of philosophical orientations together, rather than merely to take on one or two themes alone. Viveiros de Castro does this by comparing the Kantian complex to lived Amazonian practices of subjectivity and to a wider field of intersubjectivities than Kant entertained in his human exceptionalism.

A key lived difference between Euro-Kantians and Amazonians, to Viveiros de Castro, is that Kant invokes human exceptionalism tied to the postulate of a single god, while Amazonians inhabit a world in which diverse animals, plants, gods, rivers, ancestors, and spirits, as intentional subjects, enter into complex relations with one another and humans. The jaguar, for instance, perceives human blood as delicious beer to drink, while humans perceive the jaguar to be a danger. Two different perspectives intersecting at key moments. As Viveiros de Castro puts it, according to Amazonians (though specific Amazonian peoples differ in details) "the way humans perceive animals and other subjectivities that inhabit the world—gods, spirits, the dead, inhabitants of other cosmic levels, meteorological phenomena, occasionally even objects and artifacts—differs profoundly from the way in which these beings see humans and see themselves."[22]

Amazonian culture, then, does not oppose Kantian subjectivity by reducing human subjects to a mode of objectivity, the critical strategy adopted by reductive materialists in Europe. It challenges it by *multiplying lived subjectivities* and exploring how multifarious relations between them may be established. The *Shaman* helps human participants to deepen awareness of jaguar and spirit subjectivities. The priest, the therapist, and the ethologist help Europeans to pursue crossings of their own.

Modern western ethology, indeed, has finally moved closer to this Amazonian perspective on at least this front, identifying highly diverse subjectivities and cultures among bacteria, whales, plants, forests, fungi, wolves, octopi, as well as identifying variable and subtle relations across these entities. Finally puncturing the Augustinian/Kantian exceptionalism that has held the high ground in Euro-America for so long. During the long trance of Euro-American anthropocentrism, people sought to avoid the dreaded "anthropomorphic fallacy"—the fallacy of acting as if any beings beyond humans possess complex intentions, relations, and strivings. So the Amazonian challenge could become a game changer.

But what, some will counter immediately, about those spirits, ghosts, and ancestors personified by Amazonians? Must we now accept them, too, as subjectivities in order to commune with—that is, to think with—Amazonian multiplicity?

The way I respond to the Viveiros de Castro account, at least, we are now spurred to continue thinking, comparing, and experimenting in more horizontal ways with such nonwestern ways of being that may well have important things to teach us. For example, as I begin to appreciate the human self as a multiplicity (more than a unity), as I note how purposive drives become installed in the self as micro-agents that surge into each other in response to new events and then filter imperfectly into conscious life, as I appreciate more acutely how dead parents, grandparents, coaches, teachers, priests, pets, lovers, movie stars, union leaders, and politicians inhabit multifarious drives that help to constitute my subjectivity, I can now enter into more reflective communication and exchange with the cultural orientations Viveiros de Castro outlines, allowing them to work upon my thinking even as I may make provisional modifications in them.

The Amazonian account encourages more of us to think of embodied premonitions and prompts in our thinking as something like ghosts and spirits operating within us as subtle efficacies. We may be prompted, today, for instance, by ghosts we have absorbed from the past who espouse planetary gradualism, white supremacy, and exceptionalism, even after we have lifted those assumptions from the more refined and conscious categories of thought. How do you exorcize some ghosts and nourish others?

Here you do not merely "bracket" your own cultural "assumptions" as you study "them"—I wonder if that strategy ever really works that well. Rather, you encourage a back-and-forth movement between disparate perspectives. Don't allow either exceptionalism or the pose of neutrality to block reciprocal explorations across cultural perspectives. And don't worry *so* much about "cultural appropriation" of the other that you cut off at the pass fertile possibilities of reciprocal illumination. This is not the time to spin cocoons around Euro-American cultures. Especially during a time when previously unexpected planetary events call into question dominant modes of thought that have governed Euro-American cultures. And, more intensely, governed the cultures of settler societies.

We, too, are inhabited by spirits and ghosts. Some need to be exorcised; others habilitated, such as, perhaps, messages from Jocasta, Thucydides, and Baldwin. Gymnastics may help here. Experimenting, say, with new role performances that, as enacted, help to dampen some old cultural residues and prompts and to institute others. For premonition, thinking, subjectivity, gymnastics, judgment, and action are interwoven.

Amazonians, as European pagans such as Hesiod, Sophocles, and Ovid had also done in their ways, challenge a series of dichotomies pursued by Descartes and, a bit less rigidly, by Kant: subject/object, nature/culture, body/soul, thinking/gymnastics, appearance/reality, and human/nonhuman divisions among them. The subject/object division, for instance, devolves for Amazonians into complex relations between multiple subjects of highly diverse sorts. Any subject, too, can become an object, as when a cougar attacks a human or when humans cut down trees, kill jaguars, or drive elephants into oblivion.

But at other times such relations become intersubjective on a broader bandwidth. Each subject initially peers and strives out of its perspective, sometimes then adjusting its conduct in response to other subjective strivings of diverse types. You might (with refined instruments) observe plants and fungi that release poisons when under attack from beetles, broadening your sense of their range and modes of agency. Westerners unalert to how tropical crocodiles enter into collective communication below the level of human audibility, for example, place themselves at risk of being pulled into that famous death roll as they stroll across a log bridge.

The nature/culture dichotomy now becomes loosened, too. Many aspects of "nature" are now experienced to be replete with cultural relations, and many beings of disparate sorts are now acknowledged to strive and intend in relation to others. So Kantian human/nonhuman relations become reworked radically.

Many Amazonians, Viveiros de Castro says, project an initial world that was composed only of humans, a world that then devolved into a plurality of living beings of different sorts. Each devolved culture is now inhabited by subjects with disparate capacities and tendencies of internality, intentionality, striving, and interpretation. While most whites in the west do not project an initial world of humanity that devolved, that is one way to address a world composed of multiple subjectivities. It is not so far from the world Ovid projected.

The Adam and Eve story starts with humans, too, but it does not devolve in this way. It perfects human exceptionality in relation to other species while also, on several readings, contaminating its own subjectivity through original sin from within. Ovid, on the other hand, does talk about humans morphing into multiple other subjectivities, as Arachne did, for instance. He seems to respect the Pythagorean idea that humans devolve into other creatures after death.

A world of emergence, as I articulate it, appreciates a world of multiple subjects in a complementary way. Proponents of this orientation can now become more attuned than adherents of either eminent causality or human exceptionalism to multifarious subjectivities experienced by Amazonians. Those who embrace a philosophy of emergence often project a world in which new multiplicities, first, emerge out of nonlife and, second, evolve into a heterogeneous variety of intersecting lives and striving perspectives. It is interesting to see how an Amazonian philosophy of devolution and that of emergent evolution touch, particularly when each is compared to those that give priority to exceptionalism and eminent causality. Both of the former orientations are more apt to spread agency and striving widely around, to, say, insects, bacteria, cougars, viruses, fungi, and dogs: devolutionists because the initial humans from which the other beings devolved had these characteristics, emergentists because all forms of life are endowed with variants of agency and striving.

Viveiros de Castro acknowledges that he is summarizing a generic Amazonian perspective; there are differences within the Amazonian world not captured by such a summary. But a key lesson for our purposes remains critical. Amazonians are not, *implicitly*, little Kantians. They do not have a little Kant implanted in their souls, to be drawn out through western tutelage. This is true, as well, of nonplatonic pagans in Europe, Aztecs, and much of the world outside Euro-American academies.[23]

Perspectivism and Quantum Theory

Perhaps the most fascinating finding that Viveiros de Castro elicits from Amazonian cultures is "perspectivism." The world in which Amazonians participate is composed of multiple, subjective perspectives of highly diverse sorts. They essentially relate and intersect, sometimes altering one another. No single perspective gathers the whole truth into itself in advance. In a world composed of diverse, intersecting perspectives, no transcendental argument is apt to be powerful enough to guarantee such an outcome.

A world of multiple perspectives solicits encounters between and across them. Some subjectively toned, striving perspectives infect others; some eat others; some inspire new thinking in others through encounters that jostle their worlds. Plants learn to sense the vibrations of approaching caterpillars and develop the capacity to secrete chemicals that repel the attack. Indeed, certain fungi elicit poisons which paralyze worms that attack them. Cross-perspectival engagements. Sometimes an encounter leads to liquidation of one species. At other times, the two forge a creative relation of symbiosis.

A world of singular human subjects facing a world of diverse objects is one in which knowledge is said to be concentrated in one set of superior beings.

That which is constituted as object is said to lack both an inside and a complexity from which to form knowledge. "We" know them; "they" do not know us.

A multiperspectival world, on the other hand, is composed by a multiplicity of intersecting subjects of highly diverse sorts. They can both be heterogeneous and intersect. Each subject strives and responds. Each proceeds from the initial perspective available to it; it then may become altered in response to encounters with heterogeneous subjectivities. A fly adjusts its flight orbit to avoid being slapped by a human. The magnetic force of the spider's web pulls a fly coasting above it into its web. Surviving flies may then learn to set a higher flight pattern next time. A world of multi-entangled, striving perspectives.

Perspectivism is not relativism, the demeaned alternative both reductive objectivists and Augustinian types project upon those who challenge the sufficiency of a world said to be composed of human subjects and objects. Western relativism is typically confined to diverse human cultures, not extended to human and nonhuman cultures. It is often presented, too, as an ethical call to leave each unified human culture alone, as if any culture *could* contain a distinct, isolable unity rather than being entangled with numerous human and nonhuman perspectives.

Perspectivism is not relativism, first of all, because any initial perspective might be troubled or pressed to adjust through its encounter with others. Each perspective strives, reaches out, and projects forward into and onto others. No perspective is thus simply "observational" or isolable. Each is also marked by selective attractions, pulls, powers of absorption, and indifferences to others. It intersects others as it hunts, or plows, or eats worms, or rakes in plankton, or releases a poison, or builds nests, or resists caterpillars, or absorbs magical potions to expand its sensitivities and capacities. The tick is attracted to heat and blood; it does not discern other modes of movement. Some of them might surprise it.

Multiperspectivism poses difficult questions about truth and time *because* it projects a multiplicitous world in which no single outward pointing perspective automatically trumps all others. No transcendent view from above, outside, or beyond automatically rules. A set of perspectives may form a new assemblage, but no assemblage arrives at a godlike position from which it can look down with supreme confidence upon everything else. The challenge perspectivism poses to a single, omnipotent god is thus similar to the one it poses to secular notions of truth as correspondence to a world of objects. For, to a perspectivist, the nonhuman world is not composed only of objects.

Multiperspectivism thus projects a world equipped with surprises. Encounters. Alaskan antelope did not anticipate a looming world of rapid climate transition. Neither did those European settlers who terraformed the new world they conquered. Both life forms had to adjust to changes. Antelope migrate. And climate wreckage also presses today upon humans in tropical regions to migrate northward and upward.

Perhaps truth as coherence ranging across multiple perspectives of diverse types now becomes an aspiration to pursue. Truth as coherence, in which the set of coherences to be pursued is far broader than that posed by classical western coherence theories confined to human exceptionalism alone.[24] But the correspondence model of truth does becomes much more difficult to embrace once the subject/object duality has been dismantled and perspectivism is given more room to roam. The desire to save the correspondence model of truth may even draw many academics to cling to the subject/object model in a world where it becomes more and more difficult to save by the day. And dangerous to try.

Perspectivists do not easily anticipate a world in which either one set of subjects masters all others or an enlarged set of perspectives becomes composed into a beautiful world of intrinsic harmonies internalized by humans. Advocates of both such Euro-American images are apt to resist perspectivism. Some will resist it with all their might and all the intellectual and institutional resources they can muster. Are they carriers of transcendental narcissism, the inordinate demand either that "we" either *must* be in charge *or* that the world *must* harbor tendencies to harmonize with our highest aspirations? Have not pagans of diverse sorts paid a huge price for western addictions to these two varieties of species and civilizational narcissism?

The Viveiros de Castro study invites mutually illuminating conversations between Amazonians and, say, Euro-American devotees of Nietzsche, Haraway, Bennett, or Whitehead, more than it does between any of them and Descartes, Kant, Newton, John Stuart Mill, or Bertrand Russell. Think of the moment in *The Will to Power,* for instance, when Nietzsche says,

> Physicists believe in a "true world" in their own fashion: a firm systematization of atoms in necessary motion. . . . But they are in error. . . . And in any case they left something out of the constellation without knowing it: precisely this necessary perspectivism by virtue of which every center of force—*and not only man*—construes all the rest of the world from its own viewpoint, i.e., measures, feels, forms, according to its own force. . . .
>
> . . . My idea is that every specific body strives to become master over all space . . . and to thrust back all that resists its extension. But it continually encounters similar efforts on the part of other bodies and ends by coming to an arrangement ("union") with those of them that are sufficiently related to it.[25]

So Amazonian perspectivism invites exchanges with several minor and pagan orientations in Euro-American thought, especially those outside the Augustine/Descartes/Kant matrix of either transcendent or transcendental insistence. But what about those western physicists whom Nietzsche noted, concentrating as he did upon Newtonians who reigned over physics in the nineteenth century? Don't they, doesn't Einstein, postulate a boundary that perspectivism cannot cross? Don't the objects of physics fit a law-like model of explanation? Don't projected engagements across subjectivities of different sorts stop there? Doesn't the west possess physics as a master science? Don't those who lack it do so at their own risk and to their own detriment?

Perhaps so. But it is fascinating to see how a contemporary quantum physicist—himself locked neither into the Kantian subject/object division nor to Newton/Cartesian/Einsteinian philosophies of time—suggests that this non-human world itself, very much including quantum processes, *is* composed of multiple, intersecting perspectives. Let's listen to a few formulations about the perspectivism advanced in *Helgoland*, a book in which Carlo Rovelli explores the implications of quantum physics for micro- and macro-processes:

> If we imagine the totality of things, we are imagining being *outside* the universe, looking at it from out there. But there is no "outside" to the totality of things. The external point of view is a point of view that does not exist. Every description of the world is from inside it. . . . The world *is* this reciprocal reflection of perspectives.
>
> Every interaction is an event.
>
> It is a world of perspectives.
>
> The long search for the "ultimate substance" in physics has passed through matter, molecules, atoms, fields, elementary particles . . . and has [now] been shipwrecked in the relational complexity of quantum field theory and general relativity.
>
> The world fractures into a play of points of view that do not admit of a univocal, global vision. It is a world of perspectives.[26]

Rovelli's perspectivism is not identical to that of the Amazonians characterized by Viveiros de Castro. Amazonians are perhaps more apt to accept what western philosophers call panexperientialism (or panpsychism), holding that *every* entity possesses at least traces of intentionality. Rovelli supports perspectivism but does not support panexperientialism. How?

He contends that intentionality, inwardness, striving, and so on, have *emerged* in the world, and they surely extend well beyond humans. Life itself was organized from vibrant resonances between diverse modes of nonlife. Species evolution got going through a dissonant conjunction between complex and disparate elements.[27] So there are numerous, heterogeneous subjectivities, but not all entities embody the spark of life. Has Rovelli's perspectivism, then, now encountered its limit?

Not quite. Rovelli also contends that each nonliving entity, when disrupted sufficiently, may enter a new zone of criticality in relation to others. It now becomes more *sensitive* and susceptible to novel *responses*. Its behavior cannot now be reduced simply to external determinism. This is when, perhaps, Rovelli's general theme becomes even more acute, the theme that it is impossible to separate "properties of an object from the interactions in which these properties manifest themselves."[28] So glaciers, ocean currents, volcanoes, hurricanes, and so on, though unfortunately ignored by Rovelli, may elicit perspectival traces at key moments. They may even arrive at new tipping points. That is, at least, how I read Rovelli's projection of a world of multiple perspectives of different types.

I certainly do not suggest that the perspectivisms of Amazonians, Nietzsche, and Rovelli coincide. They do not. For example, Nietzsche is apt to postulate more internal reserves carried over from one relation to the next than does Rovelli, so that new relations crystallize properties but do not always create them from scratch. But each of these traditions has more *affinities* to the other two than any does to Descartes, Kant, Tocqueville, Newton, John Stuart Mill, John Rawls, Hannah Arendt, or Einstein. Continued conversations across the former constellation may thus be illuminating during an age of climate wreckage.

Perspectivism in its various guises, then, poses powerful challenges to the worlds of Descartes and Kant. But what about the Kantian postulate of time? Perhaps it is timely to explore a positive alternative to it, too? Critique alone is not enough.

Fourth Coda

Nietzsche and the History of an Error

In *Twilight of the Idols,* published in 1887, Nietzsche reduces several crystallizations of western thought to a few formulae.

Why? To let us feel more palpably what kinds of existential insistences and cultural hopes in-form and constrain each formula. And to suggest how each, even at its historical peak, faced internal and external pressures to become otherwise. I don't read these mythic phases as a kind of Hegelian dialectic, where each defeats itself and sets the stage for its necessary improvement. Though many are understandably tempted to do so. Nor do I read them as a mere, condensed history of western philosophy. Each mode becomes too entrenched in cultural practices for such a confinement. No, I read each as an actual cultural crystallization in which things might have been otherwise. And each faced impressive challenges during its heyday. Moreover, semblances of each can return. Nietzsche's own appreciation of spurts of chancy conjunction and pulses of creativity in life encourages such a reading. So do his attempts to uncover culturally embodied, spiritual prompts and demands that feed into each crystallization, prompts installed by disciplinary and ritual processes.

Each crystallization is countered before and during its heyday by other perspectives that do not speak so urgently or widely to dominant existential demands of the day. In the way, say, Sophocles challenged Plato, or Spinoza and Hume challenged Kant in advance, or Marx challenged Hegel, or Prince Kropotkin challenged Marx's image of planetary gradualism.[1] None, either, founds itself on a set of ironclad arguments. Each slips existential worries, needs, and aspirations into its manifold, in ways that exceed the arguments upon which it typically purports to rest. Thinking as embodied, acculturated ways of being. Thinking as stretching and straining established perspectives as it "sounds out idols." Nietzsche's abbreviated presentations of the whole set, then, set the stage to place another cultural spirituality into competition with these, one more suited to a world of becomings than to a world of being.

Let's review the crystallizations, as they roughly repeat, with variations, the course we have followed to that point in this text. I will freelance a bit as we proceed, at once trying to remain true to the gist of Nietzsche's reading, folding points from other Nietzsche texts into it, and adding new experiments and experiences to the themes. That, I think and hope, is the kind of reader Nietzsche sought to activate.

"1. The real world, attainable to the wise, the pious, the virtuous man—he dwells in it, *he is it*."[2] This is the world of Forms articulated by Plato in the *Timaeus*, regardless of whether he himself ever completely embraced it—since it is elaborated through the voice of Timaeus, an astronomer, rather than by Plato in his own voice. The world of Forms is pursued in *The Symposium* and *Timaeus* in complementary ways, the second dialogue enunciating the logic and aesthetic of pure Forms toward which everything tends; the first preparing exceptional thinkers to be drawn mystically toward those Forms as metaphysical realities at the base and summit of everything. Ideally, political rulers will immerse themselves in this combination of mathematical beauty and spiritual uplifting.

The true philosopher, attuned to geometrical Forms as true models, recognizes good copies in this world when he encounters them; he is thus equipped to draw others toward virtues they may more dimly sense but cannot themselves recognize so sharply. Noting with Timaeus that the human head is shaped like a globe, for instance, allows human philosophers to identify the superiority of humans over other beings. For the globe, like the circle and the triangle, is a superior form. The philosopher of Forms is the best equipped to rule and to pursue hierarchical harmony in society; other levels of society must be drawn toward faith in the piety, attunement, and wisdom of such philosophers. This is possible because they, too, are equipped with faint attractions to these Forms.

On such a reading democracy is always a danger, since it encourages people to trust their own perceptions too much, and it does not sow respect to an aristocracy highly attuned to Forms. The senses, perception, experience—so variable in themselves and differing so much between constituencies and across cultures—are not to be trusted. Trust the Forms, if you have access to them, and trust the authority of others if they do so. Democratic movements are under suspicion, then, as they challenge the authority of the Form merchants.

This crystallization, hostile to the themes of a rocky planetary world advanced by thinkers such as Hesiod, Thucydides, and Sophocles in and around

Plato's day, has a hard time withstanding the collapse of the Greek City State and its conquest by the Roman Empire. Must a plague mean that the city has forsaken the Forms? What about a conquest, or a massive volcano?

"2. The real world, unattainable for the moment, but promised to the wise, the pious, the virtuous man ("to the sinner who repents")."[3] The idea of the Forms now "*becomes a woman*, it becomes Christian." The Forms become personified. They take the shape of a personal, omnipotent, salvational god, a god to be worshipped and placed at the pinnacle of authority in the Roman Empire.

When Nietzsche says it "becomes woman" he means, first, that the earliest stages of Christianity in the Gospels were inspired to a large degree by devout women. Then male priests shouldered them out of leadership posts. Later, "woman" means above all that *obedience* to divine authority, as interpreted by male priests, becomes a paramount virtue. Nietzsche's own orientation to women deserves sharp critical scrutiny, too, but let's keep our eyes fastened now on the Augustinian orientation: women as prime exemplars of devotion and obedience to god. Monica, indeed, is the ideal devout, subordinate, maternal woman, to Augustine, her son.

Augustine himself, you might say, acts like an obedient woman in relation to the god he confesses and a male ruler over other humans, demanding traditional, feminine obedience from parishioners, nuns, and monks below him. He confesses and obeys upward; they confess through his mediation to the Being above.

This is the world of Christian empire, one in which non-Platonic variants of paganism are demeaned and defeated. Why? Above all, such pagans do not worship a single, divine, personal being who created the entire world from scratch and can thus promise the real possibility of eternal life. These pagans are too earthy, too oriented to limited gods, too oriented to a planet that behaves capriciously. If you place the demand for eternal salvation at the top of your existential list, the pagan gods need to be expunged. Especially if you yourself cannot tolerate living shoulder to shoulder with people who do not embrace the one afterworldly faith that fills you with hope. The pagan gods now become "vile." Such pagans even sometimes say that human beings are not the highest concern of these gods! Omnipotence and benevolence are needed *together* to push humans to the top of the earthly heap, the imperiously devout Augustine says.

Plato was on his way to Christianity, according to Augustine, but the birth of Christ arrived too late to enable him to *personalize* the Forms.

"3. The real world, unattainable, indemonstrable, cannot be promised, but even when merely thought of a consolation, a duty, an imperative." What we now have is "the same old sun . . . grown sublime, pale, northerly, Königsbergian."[4]

We have marched into the colder territory of Kant, a world in which "apodictic recognition" of the Form of morality itself sets in motion a series of necessary, subjective *postulates* about free will, universal morality, an omnipotent god, eternal salvation, progressive time, and the impersonal cooperation of impersonal natural forces in promoting it. What was once known by Augustine directly, or at least confessed devoutly through authoritative experience and sacred scripture, now becomes more oblique, circuitous, pale.

Universal morality now precedes theology. But the same god shines through the postulates, once the priority of morality is assumed. And the more you repeat the postulates, the more those "gymnastics" become installed in the operational premonitions of life. Becoming more than postulates.

Given the character of the Kantian world, you must postulate a divine thing in itself, an intelligent being who grasps and makes possible that which exceeds finite human understanding. You can say no more about it, though, even if the postulate secures willful freedom, human exceptionalism, and the possibility of salvation. You can, then, *think* the position and necessity of the thing in itself, but say nothing more about its characteristics. Augustinianism becomes pale, sublime, northerly, Königsbergian.

Universal morality now sings. It is now time for Europeans to come to terms with the singular course of world history, with different races and civilizations ranked in a historical hierarchy along a trajectory of universal progress. The most advanced civilizations must now tutor the less advanced. To be a higher civilization—one that grows out of the earlier, more direct Christian faith—is to acknowledge the responsibility to help lower races and civilizations climb toward the summit. No civilization is at the summit, but a few in Europe are close to it.

Kant seeks to secure this universal philosophy of morality, faith, and time through a series of tight transcendental arguments. That system reduces the sphere of knowledge to secure the reliability of its appearances, but these reductions now begin to expose the fragility of the system. What, for instance, to do with or to those who refuse to recognize the apodictic form of morality? Are they really "mad" and "raving," as Kant says from time to time? What about those, distributed widely in pre-Christian Europe and nonwestern cultures, who refuse the subject/object dichotomy, who break with human exceptionalism, and/or who experience a world of multiple lived, interacting subjectivities of different sorts? What about those who counter Kant's necessary postulate of human progress with a sense of the periodic volatility of the nonhuman world? What about those who entertain all such adventures? You can't, for instance, later insert the emergence of life from below and evolution into the Kantian system and then leave everything else intact. Such intrusions buckle the system. To Nietzsche, at least, "the will to system" reveals "a lack of integrity."

4 and 5. I now compress the next two crystallizations—or lived cultural temptations—into a larger story, doing so to allow Nietzsche to interrogate recurrent temptations to variants of *logical* positivism in advance. We arrive, then, at "the cock-crow of positivism." Here the thing in itself is no longer wanted or needed. Life becomes more "cheerful." A bit like the world Jocasta sought in the early going of *Oedipus*, perhaps. All you need now is brute data plus logic, joined to the assumption that we basically inhabit a world of solids. Positivists either forget or demean indigenous cultures. Wetlands and bogs are also too messy and disorderly to be prime objects of explanation. They are ripe to be transformed into parcels of farmland or cities of concrete.

One set of Euro-positivists, led by Auguste Comte, streamlines and simplifies the things of this world until they become suitable enough for authoritative explanation. Another movement of positivists, exemplified by Jeremy Bentham and John Stuart Mill, is eventually sharpened and consolidated by Bertrand Russell, Hans Reichenbach, and A. J. Ayer into logical positivism. They project a world composed at its base of brute data. A brute datum is what it is and nothing else. If you, first, identify such data as in themselves unsusceptible to multiple meanings and variations, and, second, compose a binary logic that separates each datum neatly from others, you can begin the long project of constructing a definitive science of society to commune with the (putative) definitive, law-like science of nature nearing completion. The world of objects is now set in a time of homogeneous instants, or as later characterized, particles and photons.

It's time, then, to be cheerful, to forge a science that enables elites to deploy the laws of society to regulate nature and improve behavior. To such projects, in broad outline, Nietzsche poses several objections. The beings who pursue them "blink too much"; they blink over the overlaps, spillovers, creative conjunctions, and evolutionary surprises marking the world. Since positivism is making a belligerent comeback once again in the social sciences, perhaps we should review its characteristics more closely.

First, the new science elides human freedom; above all, it subtracts pulses of real creativity from the human world, aspects of the nonhuman world, and multiple entanglements between them. For freedom, to Nietzsche, is not reduced merely to *choosing* between two established options. It also involves participation, in relation to other forces and agencies, in *pulses of creativity*. Positivists are imbued with drives to make everything and everyone, at least below the explaining elites, into "metronomes."

Second, positivism invests—as if it reflects a law of being itself—a human formation of a binary logic with overweaning authority. A binary, or two-valued, logic insists that every entity is either this or that and that every statement is either true or false. The science of positivism is not grounded in experience, then, but in "data" formed by forcing the vagaries of experience into these logical artifices.

Third, positivists claim that every philosophy or science denying these first two postulates resides in fatal contradiction with itself. Positivism thus first worships a divine logic of its own making and then convicts those who would transcend it of contradicting themselves. No wonder its proponents are so cheerful. And such relentless simplifiers. And, so often, authoritarian.

But such judgments lack the aura of necessity they claim to convey. No "brute datum" is merely what it seems to be *now*. Every entity, and every-thing, *bristles*, in its internality and through relational entanglements, with more than it is. Even protoplasm, Nietzsche already says in the nineteenth century, *strives* beyond itself; it also *absorbs* alien elements beyond its need for mere survival that then help to change what it is.[5] In fact, it is illuminating to trace the historical migration of the term "brute" across social endeavors, doing so to glimpse the subjugations and tyrannies that can arise from forcing facts into such frames. Descartes concluded, for instance, that animals are "thoughtless brutes," making them perfect objects for human use. Europeans who enslaved Africans and conquered the Aztecs often identified the people they subjugated as brutes. Such cultural associations slumber in the background of the twentieth-century positivist idea of a "brute datum"; the latter evacuates all liveliness and pluripotentiality from that which is so construed. Positivists seek brute data upon which, through logical extrapolation, tight explanatory systems can be constructed. Why? What spirituality drives that effort?

Logical positivists, as well as the rational choice theorists and neoliberals of today who mimic much in them, are drawn spiritually to modes of social control and regimentation; these philosophies of social explanation are invested in control, mastery, and subjugation of the "data" they study. That is why they are so intense about protecting this philosophy. They also unconsciously carry forward the assumption of planetary gradualism, one of the most dangerous assumptions within modernity itself. Nietzsche was clearly at odds with *that* assumption too. The ideas of brute data and gradualism lend support to one another; for how brute could any datum be if it and its environment periodically tend to self-reorganization?

A positivist philosophy of science is not, then, merely a way to *represent* phenomena—atoms, animals, Aztecs, pagans, cultures, forests, protoplasm, genes, and viruses. It is also a workbook on how to *organize* and treat those representations. *To make them become what you represent them to be.* That is the spirituality of authoritarianism lurking in positivism. Think merely of how compulsively detached positivists of today in the social sciences seek to discipline graduate students until that philosophy of science permeates the pores of their being.

Recently, however, molecules, atoms, viruses, planets, black holes, rational choice actors, solar systems, trees, Blacks, women, and discrete voters have lost or thrown off the appearance of unitary isness. The "is" becomes both what appears in its dominant mode to specific observers and relational pluripotentialities festering in those appearances. Singularity thus collapses into temporal entanglements. Think about viral crossings, DNA intrusions into nonnucleated cells that instigated species evolution, the rocky history of the ocean conveyor, racially selected voters forced to wait in line without food or drink for hours, trees tied together by complex root intertwinings and fungal networks, plants that release defensive chemicals when they sense the vibrations of approaching caterpillars, and climate-related asymmetries that mark the earth's path around the sun in this solar system.

We have already noted how the contemporary quantum theorist Carlo Rovelli supports perspectivism, a Nietzschean theme. He adds a point that fits Nietzsche too: "The long search for the 'ultimate substance' in physics has passed through matter, molecules, atoms, fields, elementary particles . . . and has been shipwrecked in the relational complexity of quantum field theory and general relativity."[6] The upshot of the findings of Nietzsche and Rovelli for logic are powerful. According to Nietzsche, a binary logic "is tied to the condition: assuming *that identical cases exist*. . . . In order to think and conclude logically, the fulfilment of this condition *must* first be feigned. That is: the will to *logical truth* cannot realize itself until a fundamental *falsification* of all events has been undertaken."[7] It is thus not the logic alone Nietzsche castigates, but the pretense that it fits the world like a well-worn glove. Alfred North Whitehead makes a similar point when he says that two plus three blocks equals five while two plus three drops of water might equal one.

That does not mean, though, that the mode of interpretation Nietzsche embraces is antilogical. It means, rather, that *binary* logics do not square either with the element of creativity in thought and being or with the relational abundances of intersecting entities and processes. Just as the assumption of gender binary occludes modes of being overflowing both poles. Such a logic both suppresses cultivation of new modes of responsiveness to things and depresses heightened awareness of becoming. To appreciate responsiveness and creativity—both as they emerge from the self as multiplicity and as they surge through dissonant conjunctions between heterogeneous agencies and forces—is to exceed the prison house of binary logic.[8] And it is to seek looser coherences among the multiple elements that make up a philosophy or theory.

Positivism, in its various guises, has waxed and waned in Euro-America, at least from the middle of the nineteenth century to today. Thinkers such as William James, Henri Bergson, Alfred North Whitehead, Bruno Latour, Brian Massumi, and Donna Haraway, prompted by the demise of the Newtonian system, pushed positivist orientations into a defensive position within universities during both the early and close of the twentieth century. Between these two events—particularly after Einstein's famous "defeat" of Bergson in their 1922 debate on life and time—logical positivism entered its heyday. The 1930s version in Vienna helped practitioners evade both Christian authority and fascist incursions. Was it, indeed, an (understandable) academic mode of escapism, one that nonetheless anticipated its own modes of control?

Findings, however, in quantum theory, bacteriology, virology, horizontal species evolution, complexity theory, ecology, and geology once again placed various positivisms on the intellectual defensive. Today, nonetheless, there is a new surge of positivism in the social sciences, often tethered to authoritarian university presidents and neoliberal social agendas. They often invest high hopes in artificial intelligence, though it is doubtful how many lower administrators, low-income workers, hamburger jockeys, teachers, Amazon drivers, film writers, or copyeditors share that enthusiasm.

Such shifts and turns in positivism may teach us that Nietzsche's "phases" follow no single or simple trajectory of time. When drives to mastery over people and the earth intensify, some version of positivism tends to surge in Euro-American social sciences; when they decelerate, alternative modes may begin to rise. It is not that the first forces determine the second, though, but that such pressures play roles in the struggles of intellectual life. The recent proliferation of neoliberal think tanks in America—eagerly funded by superrich neoliberals eager to protect their wealth and supremacy—underlines the return of such demands. Neoliberalism, rational choice theory, structural functionalism, closed system theories are merely some of the forms new positivist orientations assume today. They are agents of metaphysical simplification.

Today, when capitalism struggles to maintain itself against multiple political and planetary counterpressures, you once again find a class of all too confident positivists eager to apply imperious models to other peoples, other species, undercapitalized regions, mundane work life, the ethos of university life, and the earth. Neoliberal university presidents and trustees take the lead. They don't win the arguments; they shape institutions and cadres in which those arguments are housed. When positivists say their sciences are "value free," they really seek to place the metaphysics of brute facts, binary logic, and mastery above tolerable modes of contestation. In the name of precision. One-liners often suffice to subdue the opposition.

Finally, it may be noteworthy that 1930s positivism and new semblances of it today differ in dominant ideological tones. Early versions, expressed by Carnap and Russell, were leftist, pacifist, and egalitarian. Later versions, more neoliberal. Linking those differences together, however, is an orientation to social control and mastery over the earth. We can see it expressed, for instance, in this 1925 statement by Bertrand Russell. "Abstraction, difficult as it is, is the source of practical power. . . . A financier whose dealings with the world are more abstract . . . is the source of practical power. . . . Similarly, the physicist who knows nothing of matter except certain laws of its movement, nevertheless knows enough to be able to manipulate it."[9] This drive to mastery over the earth through abstract sciences, lodged in abstract models of data and logic, and inspired by one model of physics, we may today discern, is one of the drives that helped shield a class of scientists from discerning the time of climate wreckage. Once multiple sciences between physics and the humanities unfold, as they have today, set in diverse, interacting temporalities, *that* myth of science, temporal singularity, and control rings hollow.

"6. We have abolished the real world: what world is left? the apparent world perhaps? . . . But no! *with the real world we have also abolished the apparent world!* . . . Mid-day; moment of the shortest shadow . . . INCIPIT ZARATHUSTRA."[10]

This last, aspirational formulation has sometimes given trouble to Nietzsche interpreters. Are there not even stable *appearances* to confront, entertain, examine, explain, evaluate, or change? Is there, then, no-*thing* left? Has the world been reduced to a chaotic rush of accelerating temporalities—the kind of world Angela Carter portrays in *The Infernal Desire Machines of Doctor Hoffman* when the evil doctor speeds up time so much that our desires disappear before we can enact them and threats faced move too fast to be addressed?[11] No. Though tipping points arise periodically that approach such extreme speeds in specific zones. The chaos reading of phase 6 reflects the limited imagination of those stuck in the metronome logic in phases 4 and 5: "If the world is not fixed, it must be chaotic." Binary logic. Such theorists need to be put out to pasture before it is too late.

The way we have read phases 4 and 5—even proceeding beyond the time of Nietzsche—helps us to approach phase 6 as inspirational possibility. The "real world"—projected as either a permanent god or brute data—now morphs into a world of multiple, periodically intersecting, *becomings* moving at different paces. Things form and structures persist, but no-thing is permanent, let alone eternal: neither a god, nor the solar system, nor a brute datum, nor a planet, nor the ocean conveyor system, nor a steady glacier flow, nor the shape of continents, nor a religion, nor human life, nor dinosaurs, nor capitalism, nor a viral formation. Every-thing, replete with incipient potentialities, is entangled with other modes of becoming, some set on long, slow change, others set in faster metamorphoses that may periodically accelerate. It is a world of *emergences*, in which life itself has emerged from nonlife. And a world of creative evolution, in which each emergent species, through multiple encounters and entanglements with others, evolves into new modes and forms or goes extinct. It is a world of horizontal gene transfers, as well as evolution through the slower processes of mutation. (Nietzsche thought that Darwin's theory of evolution downplayed too much the role of the internal strivings of species in the evolutionary process.)

This all means, again, that no fact or being simply "is"; every isness—every fact—is inhabited by incipient potentialities that could be activated by this or that encounter. Even old cars rust. Glacier ice liquefies. A settled landscape can be convulsed by a volcano. It is not that everything can become anything; it is, rather, that anything might enter into new, surprising entanglements. Current circumstances have encouraged formation of a new species out of grizzlies and polar bears, for instance.

Can more of us come to accept such a world of becomings? (Nietzsche, as we shall see, does not demand that *everyone* do so. He leaves the demand for universal acceptance to Augustine, Kant, and positivists, in their various guises.) Two preliminaries are particularly critical here. To invest affirmation of a world of becomings and extinctions more deeply into the self and larger constituencies, more of us must work harder to revise our tacit orientations to death. Affirming death as a condition of life. More of us, too, must come to appreciate the shaky place of the human estate on a periodically turbulent planet. A "free death" would be one in which you strive to sink acceptance of death as a condition of life more deeply into the sensorium, as you also lift it into more refined regions of thought, coming to embrace, if you can, a more profound appreciation of death as a condition of life itself, seeking if your demise makes it possible meetings with friends and relatives to celebrate life. Much in your body will rebel against such an effort, certainly, on both the visceral and refined registers of being. But to embrace a free death, *if* you *can* do so when the time is right, is to express love of an essential precondition of life itself. For life secretes death, always, and death fertilizes new life. Such efforts, to the extent they succeed, help to clear the cultural atmosphere of existential resentment, that is, of *ressentiment*.

More, though, is needed. It is time to cultivate—and to bring these tidings to others—appreciation of the periodic shakiness of species life on the planet that houses them. For periodic volatility and creativity are two sides of the same process. "I beseech you, my brothers, remain faithful to the earth,"[12] says Zarathustra. In pursuing "love of the earth," Z projects *neither* a mechanistic world highly susceptible to human mastery *nor* a harmonious, organic world to which we must become smoothly attuned. Neither/nor. Those two contending idols are contested together. The hegemony of the cultural debate between them blunts thought about the planetary condition, as we saw in chapter 3 when Todorov almost automatically contrasted the new imperial world of conquest and mastery to a putative old world of harmonies projected both into the European past and a preconquest Aztec world.

The existential effort now is to overcome the deep cultural disappointment—labeled by Zarathustra *ressentiment*—that arises when *neither* side in this long-term, Euro-American debate seems to fit the world. Indeed, it never did. To allow that initial disappointment to sink in too deeply means that you and your allies will continue to express resentment against all those who present a more precarious and wondrous world of becomings. Thus Spoke Zarathustra.

Zarathustra, and Nietzsche, too, are aware that the world they seek to embrace will be resisted with intense and ruthless fervor by many. Even by aspects of themselves. They also acknowledge—even insist—that the counter-metaphysical assumptions they confess and profess are contestable. They have not been proven, even if rapid climate heating set on several interacting registers provides them with suggestive support. So, they do not demand that this vision be embraced universally; they only hope that a larger portion of people in several positions and societies do so and that more others will accept this as a noble world view with which to enter into appreciative negotiation and reciprocal contestation. The need for a new *spiritualization* of nonviolent contestation is stated transparently in *Twilight of the Idols*:

> The spiritualization of sensuality is called *love*: it is a great triumph over Christianity. A further triumph is our spiritualization of *enmity*. It consists in profoundly grasping the value of having enemies: in brief, in acting and thinking in the reverse of the way in which one formerly acted and thought. The Church has at all times desired the destruction of its enemies: we, we immoralists and anti-Christians, see that it is to our advantage that the Church exist. . . . In politics, too, enmity has become much more spiritual—much more prudent, much more thoughtful, much more *forbearing*.[13]

"Spiritual, prudent, thoughtful, forbearing." Nietzsche, the fervent advocate of a world of diverse emergences and becomings, has become a deep pluralist. He wants a plurality of faiths to enter the public realm, as long as *each* appreciates more than heretofore others as worthy competitors. You move through new Christian and Jewish appreciations of each other, for instance, to place another viable candidate onto the field of debate, too. A *spiritualization* of enmity. Nietzsche resists every closed universal; he would thus fight against fascist movements of total control. He supports, rather, a plurality of lived, final faiths touching and disturbing one another. This is what he means by "nobility," a situation in which each nobility requires a series of other nobilities to be itself.

Nietzsche's gesture toward deep pluralism—one that does not force all heterodox faiths into "the private realm" alone—would, however, become more *multidimensional* if it also emphasized a plurality of orientations to gender, racial variety, a diversity of sexualities, and so forth. Nietzsche falls short there.

Make no doubt about it. The deepest "conjectures" of Zarathustra are not presented as certainties or unquestionable dogmas. They are presented as lived conjectures to be defended and enacted in relation to other such conjectures, if and when proponents of the latter do not close things down. Closure is the kind of world Plato, Augustine, Kant, and positivists pursued, in different keys. Even Timaeus granted his philosophy was only the most probable. The vaunted fallibilism of (the slightly revised) positivism of today is set in a larger world view not to be interrogated. Nietzsche, on the other hand, would appreciate contemporary Christians such as Catherine Keller, Pope Frances, and Charles Taylor who welcome a world marked by the spiritualization of enmity. And then he would debate each sharply.

Nietzsche, the ironic evangelist of nontheism, emergence, becoming, a volatile planet, and spiritual attachment to a bumpy earth, seeks to inspire more people to embrace such orientations to life and the earth.[14] His informed guess, in the nineteenth century, was that life would become even more precarious and violent on earth if that did not happen. He even made it clear that neither capitalism, as he then knew it, nor socialism, as he also then knew it, was up to the task. They both demand too much productivity, too much economic regimentation, too much consumer abundance, too much mastery over the earth. He could have, but did not, add to his list of things to overcome that they demand too much exploitation and domination by temperate zone, productive regions over tropical and polar regions of the earth. He was, in that respect, all too European.

There is thus much to overcome, in the aristocratic Nietzsche. His opposition to democracy (except for one brief foray), his orientations to women and race, and his reticence to address Euro-imperialism closely are three of them. Is it timely, nonetheless, for more of us to make these corrections while *also* entertaining a world of multiple becomings, a world in which death is acknowledged to set a condition of life itself, a world in which more of us admit the shaky place on earth of the human estate and other species with which it is entangled? Is it timely for more of us to struggle with those dissonant conjunctions between capitalism and planetary volatility, while at once striving to appreciate more profoundly than heretofore how the very bumpiness of the planet allows life to be and to become? Would that help in the struggle with difficult times ahead, during a period of intense planetary heating issuing first and foremost from the hegemony of extractive capitalism?

5
Amitav Ghosh, Michel Serres, and the Time of Climate Wreckage

In this closing chapter I place into conversation the Bengali novelist and anthropologist Amitav Ghosh, the Irish geologist Bill McGuire, and the post-pagan French philosopher of time, Michel Serres. Doing so, first, because these travels across fields better help us to discern the effects of two pagan conquests and the seeding of climate wreckage. Doing so, second, because they illuminate how serious the time of climate wreckage has become and help us to appreciate how impersonal planetary processes distribute the effects unevenly across classes and regions of the world. Doing so, third, because it is timely to rework images of time embedded in western Christianity, imperialism, capitalism, and secularism. The idea is that these thinkers, coming from different places and fields, may need infusions from each other to come to grips with today. And we need all three of them.

The Curse of the Nutmeg

I like nutmeg. Its pungent smells and flavors have graced the spinach, eggnog, and coffee I consume. I visited nutmeg and cinnamon trees on the Caribbean island of Grenada in the 1990s, fascinated with the bark and seeds of those trees and the pungent forest odors emanating from them. That is, I knew little about the history or powers of nutmeg and was only aware of the western imperial adventures it had excited in the Caribbean. Unalert, too, to how a unique combination of volcanoes, climate, soil, seed fragrances, imperialism, healing powers of the nut, and European stories about its ability to cure the plague formed key elements in the Dutch-Banda islands conjunctures of the south

Indian Ocean of the 1620s. There are only a few places where nutmeg can grow and exercise its potent attractions to humans.

Amitav Ghosh starts *The Nutmeg's Curse* with an event in 1621 on a Banda island, located in the Indian Ocean south of Indonesia. The island of Lonthor. Dutch colonizers invaded the island that year, intent on eliminating the local population to establish a monopoly over the flourishing nutmeg trade. Europeans loved nutmeg for its fragrance and its reputed capacity to cure the plague. The conjunction of a warm climate and volcanic eruptions created a unique setting for the spontaneous growth of nutmeg trees.

There was no central authority with whom the Dutch invaders could negotiate (on their terms); so extermination became the strategy of choice for the relentless invaders. The Dutch wanted control of the trade, and their traditions of Christian and imperial entitlements made many conclude that it was worth extermination of locals to secure it. One Dutch invader started the war prematurely when he interpreted the noise and fire created by a falling lamp to be the beginning of a native attack. The invaders were not only haunted by their ambitions but by ghosts of their own making.

What motivated these invaders? They wanted, first of all, to monopolize the nutmeg trade. Their images of the population of the island dwellers as inferior beings unworthy of respect made a major contribution. An image of nature as a deposit of resources for imperial use spurred them on too. Ghosh notes that these three interwoven existential orientations had been supported by the European philosophies of Francis Bacon and René Descartes. Yes.

I also suspect that John Calvin (1509–1564)—whose strict Puritan doctrine had by then traveled to Holland and soon traveled to North America—also played a role. Calvin carried several themes from Augustine, whom he admired, to new extremes. He radicalized the theology of predestination to honor god's omnipotence, whereas Augustine had kept the balance between grace and will murkier in the quest for eternal life. How could a god be omnipotent, Calvin said, if people could influence it by *earning* their way to heaven? He also purged signs of life and liveliness from nonhuman nature so that Augustine's Christian dominion over it could become domination. Additonally, you can't really honor divine omnipotence if the world is said to be cluttered by other life-forms not predesigned to honor you as the singular creative agent. Calvin fostered intensive church disciplines (gymnastics) to adjust and fit the behavior of its members to this doctrine. He construed pagans to be degraded beings because of their incapacities to acknowledge omnipotence and predestination and their worship of idols. But he concluded that while humans cannot *earn* salvation through success in worldly life (that, again, would qualify omnipotence), worldly success can be taken as a hopeful *sign* that you are chosen. That meant, too, that worldly failure became a sign of being doomed to hell. Weber, indeed, later translated these last themes into a story about how the resulting cultural ethos of intense motivations to succeed helped to promote the advance of industrial capitalism in the northern European states of England and Holland. The Puritan (Calvinistic) population of American settler colonialists soon followed similar strictures.[1]

All these themes—an image of inert nature, the omnipotence of god, the superiority of Christian Europe, the brutishness of non-Europeans, the importance of specific cultural disciplines, the primacy of private property, and drives to conquer nature—are repeatedly reenacted in various ways by western settler societies, as the remainder of the Ghosh book records. The cultural elements that supported colonization and conquest brought untold suffering to indigenous people in the Banda Islands, India, and the Americas; it also wreaked terrible harms on other peoples in Asia and Africa. What's more, these drives set the stage for emergence of a new time of climate wreckage, accompanied by ignorances, denials, and callousness flowing from the European and American regimes which launched it.

One theme Ghosh dramatizes is how the defeat of indigenous peoples "muted" vitalistic orientations to life. The victors carried "mechanomorphism" with them, as they enacted genocide. And as they "terraformed" the lands taken over. Terraforming begins by transforming forests, within which indigenous people have forged their lives and through which rough carbon balances had been maintained, into parcels of privately owned land suitable for intensive agriculture. The new parcels were adjoined to areas where cattle and pigs ran freely, destroying multiple species' life habitats and ravaging lands of Amerindians. Early terraforming very much helped set the stage for the Anthropocene. It became even more monstrous when it morphed into huge coal mines, concrete buildings, oil fields, offline oil drilling, and, later yet, massive fracking fields. Concrete soon became the building material of choice, spewing pollutants into the atmosphere and contributing to climate heating.

Ghosh does not think that any single "factor"—such as capitalism, or the massive carbon infusions of imperial wars, or racism, or a protestant religious creed, or protestant disciplines, or intensive private agriculture, or a positive orientation to technology, or an innate drive to mastery, or a cultural desire to protect a sanctified image of time—*suffices* to explain the western politics of expansion, conquest, and climate decimation. Rather, distinctive spiritual demands to separate the highest humans from an inert nature, imperial drives, racialization, Christian notions of superiority, capitalist intensification, and technological regimes worked back and forth upon each other to foment a new civilizational complex.

European drives to empire, for instance, precede, include, and exceed capitalism, without (at the other extreme) being detached from it or reducible to eternal, genetic drives. Capitalism is highly pertinent to these developments, then, but no *economistic* rendering of it by Left, Center, or Right suffices to capture the diverse civilizational forces that infuse and press upon it. Economism here is not confined to ignoring the spiritual dimensions that infuse regimes of profit, work life, consumption, gender, and race relations in different regimes—though that is a very important part of it. Economism also includes a refusal to commune with multiple modes of nonhuman striving and purposiveness that both populate the earth and enter into manifold relations with human civilizations.[2]

Here is one way Ghosh reviews connections between what I have called two pagan conquests and the emergence of climate wreckage: "Uncanny indeed are the similarities between the current planetary crisis and the environmental disruptions that destroyed the life-worlds of innumerable Amerindians and Australian peoples." And today, he sees, "the western settler-colonial culture is no longer confined to the settler colonies."[3] China, India, Indonesia, and Brazil, once objects of western expansion, have now joined the climate wreckage club as they strive to catch up with the earlier carries of wreckage. He could easily add Russia to that list.

Something that *almost* dies during these conquests are lived, "vitalist" orientations to nature, orientations that might have informed European practices if they had been appreciated. Here are a few things Ghosh says about this critical issue:

> [Of Davi Kopenawa, the Amazonian who retrained himself to master shaman techniques and communicate with nonhuman agencies:] Kopenawa was able to enter a realm where *all* beings are spirits—humans, animals, water, plants. . . . He is one of the few living people who can give us a sense of how the planetary crisis appears when viewed with nonhuman eyes. (208)
>
> Is a "politics of vitality" at all possible, or desirable, at this advanced stage of the planetary crisis? After all, the very idea of nonhuman vitality could be regarded as dangerous by those who believe that humans are the only beings endowed with souls, minds, language, and agency . . . , to think otherwise is to be irrational. (222)
>
> African American traditions of resistance also draw upon resources that are completely different from those of official modernity. (236)

> A necessary first step . . . [is] a narrative of humility in which humans acknowledge their mutual dependence not just on each other, but on "all our relatives." (242)

> [Of the contemporary Dutch writer, Louis Couperus:] In both novels the vitality of the setting manifests itself in the domain of the uncanny, as the frontier where colonial power meets and limit beyond which lies something unfathomable, an abyss that cannot be made to submit to the colonialist's will to impose order. (251)[4]

The proliferation of witch trials in seventeenth-century Europe and North America occurred when the Little Ice Age was wreaking havoc on both regions and religious struggles had intensified in Europe. Did the trials, indeed, express perverted awareness within Europe of nonhuman modes of agency, a sense officially denied but selectively affirmed and then concentrated only on those accused of witchcraft? The return of the repressed? Witches were said to communicate with demons (as Augustine had said pagans did); these communications, while recognized by the accusers, were defined so that they could only hurt or damage the keystones of western civilization. They could never help. Freud, if he had himself not been an exceptionalist, could have had a field day with this combination.

Pagan European and nonwestern vitalists, though, not only extend intelligence, agency, and striving to nonhuman animals, unconscious human spiritualities, and plants. They also often bestow spiritual agencies upon ancestors, rivers, volcanoes, earthquakes, floods, mountains, and deserts. How can ecological Euro-Americans, who have only recently come to terms with multiple modes of nonhuman agency in animals, bacteria, and plants; who admire the collective agency of honeybees, the wisdom of blue whales, the unfamiliar intelligence of an octopus with multiple brains distributed across several arms, the thoughtfulness of crows who sacrifice and innovate to rear their young, even the creativity that may emerge when a virion resonates with the cell of another species until a viral species crossing occurs—how can those who have gone this far at least *begin* to commune in more reciprocal ways with indigenous leaders who identify spiritual agencies in yet other nonhuman processes?

Ghosh thinks this is a critical question. I concur. Here, perhaps, is one preliminary way to push the western envelope a bit further, nudging it toward closer engagement with shamanism without necessarily drawing the two orientations into complete identity. Thus, if you construe human drives to be unconscious, purposive, propulsive, and memory-soaked formations often activated when this or that event with affinities to past experience occurs, you may now also appreciate how such activations carry ancestral influences of parents and grandparents into your and our purposive drive structures. When, say, a natural event shocks or surprises people, some of these slumbering drives may be activated. Moreover, deceased colleagues, friends, and authors may speak to you under this or that circumstance. For the self, as it ages, encompasses a large multiplicity of voices, shifting in weight as new events arrive. You now place ancestral wisdom (as well as problematic prompts) into communication with nonwestern experiences of vitality, even though the two do not converge entirely, leaving open the possibility that each tradition may further illuminate the other.

If shamans cultivate skills to sensitize themselves to new, nonhuman experiences, others can strive to do so too, as we become more attuned to diverse vitalities in and around us. As we, for instance, attend to those birds several years ago who flew in massive clouds out of Oklahoma just before an earthquake erupted. We, too, can be touched by ancestors who haunt our drives, by other species, and by whirling rivers—for good and bad. I tend to think of Kant, for instance, as a zombie who continues to stalk western culture.

Have you ever been drawn, ever so slightly, by a fast-paced river to give yourself to it? If so, that pull cannot be reduced entirely to a *metaphor* of your desire. It takes a meeting of desire and the river's lure together to instigate the pull. River gods and goddesses dramatize such lures. Such spectral attractions can perhaps help to teach westerners a bit more about the role gods and goddesses play in the lives of many peoples.[5] And about our own need to cultivate modes of sensitivity to lures that extend beyond those modes of reductive subjectivism that rather easily capture us.

Moreover, themes similar to Amerindian communication with nonhumans can be found in Hesiod, Sophocles, Dionysus, and Ovid, also opening doors to communications across time and space between European and non-European paganisms. Tiresias, for instance, was a shaman in Thebes. Did those birds he heard yelping so urgently suffer from the plague also terrorizing people in Thebes?

It took both the first and second conquests of paganism to mute voices many of us once again strain to hear. It helps to recall, too, how western neurosciences are now more attuned than heretofore to cortical, bacterial, and viral energies in the gut that maintain close communication with the multicentered brain in the head. Such purposive mini-agencies "percolate" into the head, making a difference to behavior below conscious awareness.

Or consider how recent turns in the sciences of tectonic plates, volcanoes, and earthquakes attend to earth-memory. A fault line, created by a former quake, becomes an earth-memory that helps to set the locus for a future quake. Don't build your house on a fault line. Or the edge of a dry riverbed may be activated during the next flood. Or consider how resonances back and forth between the Mauna Loa volcano and the Kīlauea volcano on the Big Island disclose how one spurs the other into action. I walked upon the lava fields of Mauna Loa a decade ago, feeling the heat vents and amazed to see new green shoots pushing up on the hard lava field, so that soon the new mounds of earth spawned would allow yet more lush growths, as they indeed had already done on a sector just off the most recent field. The new sciences of complexity in biology, bacteriology, virology, mind-gut relations, and birdlife encourage such cross-regional explorations, as do other planetary sciences that have broken with Euro-American nineteenth- and twentieth-century stories of an inert earth and planetary gradualism.

One task may be to move back and forth between indigenous spiritual orientations crushed violently by settler colonial conquests and production of climate wreckage by settler societies, doing so in part to amplify exchanges between critics in those cultures and European modes of paganism belittled by similar forces. Neither to absorb either uncritically nor to repudiate one through the pretense of undeniable cultural superiority, but to explore such experiential exchanges empathetically while simultaneously paying close heed to the contemporary world of accelerating climate destruction. I think and hope that I join Ghosh, Ned Blackhawk, Thomas van Doren, Donna Haraway, Eduardo Viveiros de Castro, Naveeda Khan, Anna Tsing, P. J. Brendese, and Jane Bennett in pursuing such exchanges.

A Hothouse Planet

Ghosh articulates a persuasive critique of mechanistic models of nature popular in the west after the seventeenth century. Moreover, he defends and explores a vitalism that extends multiple modes of creative change well beyond the human. It is not perfectly clear, however, how far he departs from a familiar temptation to embrace a harmonious image of the planetary as the best counter to the untenable theme of an inert planet. Vitalism can be pushed in several directions, as he sees when he worries about neofascist attempts to align vitalism with the singular priorities of a white nation. Another direction is toward the assumption of an organic world of natural harmonies, waiting to be approximated. But those are not the only alternatives.

I feel confident that Ghosh breaks with the organic ideal that has traditionally provided the favorite debating partner in Europe to dominant images of earth mastery. We have already witnessed how Nietzsche, Hesiod, Aztecs, Viveiros de Castro, and Keller break with both sides of that debate. It may still, however, be pertinent to fold more robustly into Ghosh's account specific understandings of how large planetary formations such as the ocean current system, trade winds, glacier flows, asteroids, monsoon interruptions, desert retreats and advances, methane storage and releases, El Niños, earthquakes, and volcanic eruptions function, particularly how they infuse and impinge upon the other agencies and forces he contemplates. Here the geologist Bill McGuire can be drawn upon to augment Ghosh, without leaving the latter's invaluable insights behind.

Before he can do so, however, McGuire himself must be purged of two tendencies Ghosh has overcome. It is not correct to say that the first source of modern climate change was triggered in England after the invention of the steam engine in the 1770s, though that event did provide it with one of its several inflection points. No, these processes were set into motion by colonial settler modes of violence against nonwestern peoples and by its reorganizations of nature. And those latter drives, themselves, were built upon the previous defeat of nonplatonic paganism within the continent of Europe itself. McGuire must also be cautioned against use of the universal "we," and "humans," words he still occasionally uses when talking about how climate wreckage was instigated and how to mitigate it, though he now veers more from those tendencies in the second of his two books on this topic. The truth is that nontemperate regions of the world have been hit earlier and harder than temperate regions—particularly settler societies in old capitalist states. And settler societies joined Europe in instigating those processes from at least the seventeenth to the late twentieth century. So I drop these two tendencies from my engagement with McGuire, as I address his augmentations to the Ghosh account. The "we" must become more selective at some moments and more invitational at others.

In an earlier book, *Waking the Giant* (2011), McGuire reviews the volatile planetary history of the earth, doing so to help explain current phenomena and to project new possibilities into the future. He knows that the nineteenth- and twentieth-century geological story of planetary gradualism inhibited earlier explorations of the current period of climate wreckage. And he knows that dominant projections from the IPCC and mainstream geologists during the late twentieth century continued to underplay the dangers and speed of change. The latter did so, first, because they sought to get favorable responses from publics, corporations, and politicians who were (and are) eager not to heed them. They did so, second, because they attended most closely to computer projections that have persistently lagged several years behind what actually was happening on the ground, within the oceans, in the air, along glacier flows, and in volatile relations between those forces. So McGuire thinks it is wise to attend closely to earlier periods when climate heating also took off, using the dynamic tipping points set into motion then to help gauge accelerations in play today.

We will confine ourselves to merely one of his reviews of earlier planetary volatility. The Eemian period started 130,000 years ago and lasted about 14,000 years. It takes its name from the Eem River in contemporary Holland. A rapid climate shift in geo-time occurred. Over a period of a few thousand years, without human aid, the landscape shifted from an ice age to a new time hotter than that commonly projected for the Anthropocene.

> The rapidity with which the climate switched from icehouse to hothouse conditions was particularly astounding, with hippos and other animals from the tropics roaming England and other northern temperature zones just a few thousand years after the landscape was buried beneath ice or frozen solid. At its peak, around 125,000 years ago, there is evidence for . . . forests reaching as far north as Norway and Canada's Baffin Island, both well inside the Arctic Circle. On average, the best guess is that global temperatures were 1–2°C [up to 3.6 degrees Fahrenheit] higher than at present.[6]

The Eemian event fostered robust glacial, climate, species, ocean, and atmospheric changes. For example, the sea rise of from 4 to 6 meters during this period (13 feet for the lower figure), shows what a turbulent planet can do on its own from time to time, even without the brutalities of settler society, deforestation, and escalating capitalist emission triggers. The Eemian provides clues about how capitalist triggers of today activate diverse planetary amplifiers that then assume a life of their own. One potent instance was how new levels of heating may have triggered massive methane releases in Antarctica during this period, setting off a new spiral of heating. Some other accounts doubt that such a methane bump occurred.[7] Nevertheless, climate change is often bound to cascading causalities, in which one process activates others and those yet others, until the accumulating effects far outpace the force of the initial triggers.

But what triggered the Eemian? That, as far as I can tell, is still partially clothed in mystery. There was a periodic eccentricity in the earth's orbit, increasing the amount of sunlight the earth received. The earth's wobble hit a more pronounced period of slant, too. Those two catalysts may have combined to set off a series of cascading events. Melting glacier ice reduced the albedo rate (reflecting level) of glaciers, setting a self-amplifying melt spiral into motion. Other things increased CO_2 emissions. This plurality of events coalesced to produce cascading effects, with each change amplifying some of the others, spawning tipping points that increased the rate of change and then circled back upon themselves to do so again. Time accelerates during such tipping points.

McGuire thus examines events in the past because they sometimes intimate ramifications that could occur in the near future. This is so because, until at least recently, computer projections into the Anthropocene future have underplayed how capitalist CO_2 emissions set such cascading processes into motion. The past is sometimes a better gauge of the future than computer projections that underplay the dynamism of intersecting amplifiers.

In *Hothouse Earth*, published in 2022, McGuire takes off his gloves. The earth has now passed beyond the time when rapid climate heating can be avoided. Much future heating, with its numerous deleterious effects, is "baked in." Multiple amplifiers, already in motion, will continue to swell and intersect, accelerating what he now calls climate wreckage.

We can add that it might have been possible to avoid this degree of wreckage if, say, fifteen or twenty years ago, all old capitalist states in the temperate zone had taken rapid, radical action to transition to solar and wind energy, rapid transit systems, electric or plug-in hybrid cars, eco-housing designs, urban bike trails, significant reforestation, circular recycling, and low-meat diets. Such regime actions, if done in egalitarian ways, might even have inspired more recent capitalist emitters outside temperate zones to adopt similar strategies. But few old capitalist states adopted radical initiatives; those who called for them were typically defined by climate denialists and casualists to be alarmists and outliers. That encouraged newer fossil extraction capitalist states—such as China, India, Brazil, and Russia—to insist that there is no call for them to take significant remedial action either.

So, the situation Ghosh characterizes is now in place, one in which old capitalist states (led by the United States) delay taking dramatic action and newer states press forward with coal and oil use. Each region blames the other for inaction, even though it must be emphasized again that old capitalist states had an opportunity earlier that few—with a couple of exceptions—took. Denmark, for instance, did so. I was impressed again, during a recent visit to Copenhagen, with its comparative ecological awareness, bike paths, walking, dietary regime, and inclusive subway system. Even there, though, the dynamic relations between capitalist triggers and planetary amplifications may be insufficiently appreciated.

The Republican Party in the United States—a key gathering point for the white evangelical/neoliberal resonance machine—continues to do everything in its power to resist both rapid economic reconfiguration at home and paying reparations to peoples in tropical and polar regions adversely affected by its legacies. The richest of the superrich plutocrats, certainly, also know what they are doing; otherwise they would not prepare obscene private escapes, such as vacation homes in northern lands, island retreats, and new space adventures.

The whole scene might remind you of a Goya painting, *Duel with Clubs* (1820), in which two antagonists fight to the death, neither noticing how the battle sinks them deeper and deeper into quicksand. Except that one protagonist is larger and more pernicious in the contemporary scenario.

Here are a few of McGuire's conclusions about the current condition, as it projects into the near future.[8] First, the oceans, key absorbers of atmospheric CO_2 during the twentieth century, are rapidly declining in their absorptive capacity as they warm and become more acidified. They could even become emitters themselves soon. Second, the Amazonian rainforest, a leading provider of oxygen and absorber of atmospheric carbon emissions, is weakening in its capacity to do either; it now "teeters on the edge of a tipping point" after which it will become an emitter rather than an absorber. Third, as global heating continues, plant life increasingly veers from being a carbon sink toward becoming a carbon source. This is in part because photosynthesis requires lower temperatures to operate most efficiently. The optimum temperature for photosynthesis is about 64 degrees Fahrenheit. Fourth, millions of years of methane storage in the tundra and ocean clathrates may later be released by rapid heating. Methane stored in the tundra and clathrates, over the short term, carries a heating potential 86 times larger than the comparable amount of carbon. Fifth, the melt-induced reduction of glacier weight over Greenland and several mountain areas increases the probable number of earthquakes, destructive avalanches, and volcanoes—which, a couple of years after erupting, balloon greenhouse gases into the atmosphere. Sixth, rapidly increasing ocean temperatures foster algae blooms, decreasing further the absorptive power of oceans and spurring another self-amplifying process. Seventh, the growing instability of polar glaciers means that oceans will definitely rise higher than previously projected, probably much higher.

For these and other reasons, McGuire says the world is undergoing climate wreckages that will accelerate and deepen for at least the near term. The task is no longer to avoid or stop processes well under way. It is to slow down amplifiers now in motion, to mitigate the worst effects baked in, and to make class and regional reparations to those most severely afflicted.

Planetary Circuits of Imperial Power

I now tie themes separately advanced by Ghosh and McGuire more closely together into a distinctive theme, doing so to heighten awareness of how the two conquests of pagans find expression today in impersonal planetary regional distributions of climatic effects. Impersonal planetary circuits distribute temperate zone capitalist regime emission triggers disproportionately to tropical and polar regions and to lower economic classes in both regions.[9]

Today capitalist deforestation, carbon emissions, and methane emissions trigger a variety of planetary amplifiers that, first, render the effects much larger than the triggers alone and, second, distribute them disproportionately to nontemperate regions that have typically fomented the fewest triggers. So geological processes now become an essential component of cultural history, theory, and the humanities.

Consider, first, how an amplifier operates in one potent but simple process. A glacier melt, triggered by CO_2 emissions that raise the atmospheric temperature, speeds up as new meltwater now absorbs much more heat from the sun than ice did. Thereby amplifying each new round of melting. A simple, potent, continuously looping and self-magnifying process. The world is full of such amplifiers. Consider how water flowing on top of a melting glacier dives through moulins (cracks or gaps in the ice) to grease the bottom of the glacier and increase the pace of its slide, another self-spiraling process.

Now we need to attend to the world distributive effects of such amplifying processes. The historical irony that old capitalist states are lodged in temperate zones means, basically, that state capitalist emissions often trigger amplifiers that then distribute the worst immediate effects into racialized, decolonizing tropical and polar zones.

Enlarged El Niños in the Pacific, for instance, reduce the absorptive power and strength of trade winds heading east. In 1897, such a conjunction between a large El Niño and weakened trade winds interrupted monsoons in the Horn of Africa and India for three years, creating devastating famines ignored by the British Empire. Africa produces about 4 percent of the world's carbon emissions today and suffers disproportionately from the destructive effects of such impersonal circuits. Impersonal, planetary circuits of imperial power, inducing huge gaps between emissions sources and the size and planetary distribution of the results. These impersonal distributions often carry class effects too, for typically the most vulnerable populations in each region—due largely to the regular dynamics of class politics in capitalist economies—live where people are most susceptible to climate ravages.

Today, such processes are underway again. Once the winds slow down on the Horn of Africa and India, other complexities set in. Some winds, for instance, then distribute rain to western Africa and parch eastern Africa. Recently, El Niño effects have interacted with the newly discovered Indian Ocean Dipole, creating patterns that manufacture huge droughts for one period followed by monstrous storms at another. Such oscillations have carried extreme consequences in East Africa, India, Pakistan, and Australia. The larger distributive effects are rather well known but I, at least, do not yet know how these latter intraregional distributions work. It is perhaps enough to say now that enlarged and long-lasting El Niños once again put eastern monsoons at risk of interruption, during a period when population densities are much larger than in the nineteenth century.

That means that western states do not merely need to make reparation and mitigation payments to regions that "happen" to suffer the most from extended drought, stormy weather, and heat waves and have the fewest resources to cope with those disasters. Western state emissions *spur* impersonal amplifiers that distribute the worst immediate effects to zones already suffering severely from other historical impositions. As these circuits become more widely understood, it will no longer be permissible to ignore the responsibilities of the triggering regimes and classes. The process is comparable, perhaps, to how early European invaders would depart an Amerindian sector before the smallpox plagues they instigated took hold. Many soon consciously implemented such scourges.

Another example. Accumulating emissions from temperate zone states may release methane sediments in the North Pole, spawning a potent planetary amplifier that renders climate heating much larger than the force of the initial triggers. This effect, if it occurs, is further augmented by the fact that climate heating weakens the polar vortex that maintains cold air circling around the North Pole. The weakened vortex now dives down periodically into temperate zone states such as the United States, producing surges of cold air there and releasing warming air surges to ramble north. These impersonal circuits provide two reasons that the North Pole is currently warming at four times the rate of temperate zone states. There are others, too, including how the tilt of the earth exerts magnified effects in the north.

The rapid meltdown of the Thwaites Glacier ice shelf in Antarctica provides yet another example. Earlier said to be at risk of collapse in seventy years or so, the shelf collapse is now projected to occur by 2030. The shelf currently holds back the larger Thwaites Glacier poised to slide down the Thwaites mountain into the South Sea after the shelf collapses. Some estimates place the additional global rise in sea level after the second collapse at two to three feet. More may well be in store too, though the issue of timing grows here. If those two collapses disturb sufficiently the huge West Antarctica Glacier to which they are attached, there will be an eventual worldwide ocean rise of ten feet. Such a rise, once set into motion, could not be stopped by human technology. Even more recent reports suggest the latter process has now passed the tipping point, though the timing of its collapse remains uncertain.

This now probable event trumps Nassim Taleb's presentations of Black Swans, which are primarily stock market events of low probability and high impact too often underplayed by statisticians. The West Antarctic Glacier collapse is a future planetary event of, say, at least medium probability and devastating impact.

This is yet another instance of triggers instigated primarily by the United States, Europe, China, Russia, Brazil, and India that place low-lying areas of Bangladesh, the coast of Egypt, Pacific islands, and Senegal at high risk, as well as ocean shores in low-lying zones of the United States, south China, and Europe. Once again, the impersonal distribution of planetary consequences instigated to a great extent elsewhere.[10] Such instigations may at first be unintentional modes of power; as they become known and continued they become intentional modes of imperial power.

Think, too, of how climate heating accelerates devastating wildfires in northern Canada, a state low on the scale of CO_2 emitters. These fires, accelerated by combustible sap that marks conifer forests, devastate the habitats of native peoples who have made no contribution to climate warming. The effects eventually recoil back on temperate zone capitalist states, too, carrying polluted air into them. And consider how the ocean conveyor which now distributes warm air to Europe and eastern United States is weakening and may be primed to stop again. The probable distributive effect would make it colder in the eastern United States and Europe while heating even more radically tropical and subtropical zones.

My final example speaks to hurricanes now increasingly common and intense in the Gulf of Mexico. The higher intensity of such hurricanes, spurred by heating water in the Gulf, has now combined with the reduced speed of trade winds that carry them, to produce longer lasting and more devastating torrents of rain in the Caribbean. Black and brown peoples in the Caribbean and southern coasts of the United States are disproportionately devastated by these storms, though the intensities and frequencies are initiated elsewhere. Poor whites are concentrated in the Bible Belt too. The slower winds, too, mean that savage rainstorms last longer, dumping huge amounts of water on regions that are themselves low emitters. Planetary circuits of imperial power.[11]

These examples show how imperialism and class exploitation both take on a distinctive character today, even as the old modes continue. What's more—and this is becoming more apparent—just as old forms of imperialism eventually returned to haunt imperial states, powerful climate recoils now haunt vulnerable zones in old capitalist states too. Wildfires, expanding drought zones, flooded ocean shores, unseasonal tornadoes, long heat waves, more intense hurricanes, atmospheric rivers, huge flash floods, periods of icy cold as weakening polar winds wander into temperate zones, and even a new halt in the ocean conveyor are among actual and potential events set into motion.[12]

Another feature of these rapid shifts must be emphasized, too. Old species niches are disturbed by rapid adjustments available to some species and delays faced by others. To note merely one example, west African penguins still return seasonally to the sea when fish they have depended upon for centuries have migrated to colder seas. Such asymmetries of timing multiply in diverse regions of the world during a time of rapid climate destruction.[13]

Old, temperate zone states today trigger processes that also drive humans living in hotter, increasingly dry and unstable regions to surge north and west in growing numbers and enhanced intensities. Temperate zone capitalist states thus instigate desperate migration flows into their own territories; they then intensify stories of climate denialism and casualism that deflect awareness of these connections by blaming the very cohorts fleeing them and other related disruptions. Such intercoded processes are apt to spiral. If and as recipient states house large constituencies who work hard to intensify intraclass racial divisions, or deny climate change, or remain casual about its magnitude, or cultivate studied ignorance about how planetary circuits of distribution and magnification work, white racist resentment against climate refugees is very apt to magnify fascist drives within old, temperate zone, capitalist states. Each of these processes, human and nonhuman, is imbricated with several others.

Catastrophism and Distributed Catastropheism

One remarkable feature of new fascist movements is that they translate old deniable racisms marking neoliberal capitalism into gleeful racism and celebrations of violence against those not counted as white, Christian, ambitious, or otherwise entitled.

In the United States, at least until recently, people like Ghosh, McGuire, me, and many others have been repeatedly warned against climate alarmism, of projecting disaster into the future unless radical actions are taken, and of being harbingers of "catastrophism." Such orientations, it has been said, play into the hands of the "populist" right, as it regularly manufactures false alarms to intensify racism and mobilize support for a white Christian nation.

Such warnings, however, are only half right, and then about only one thing. The catastrophism of the right—with its warnings against "race replacement," its Big Lies about stolen elections, its pretense that climate warming is a false flag designed by the left to justify socialism, its accusations that vaccines contain implants to track white citizen movement—these modes of catastrophism are dangerous and potent sources of right-wing mobilization. But another potent danger is how rightist modes of catastrophism *cover up and defuse attention to real, potential climate catastrophes unless radical corrective ac-*

tion is taken. The same leaders who foment catastrophism deny or occlude the prospect of climate catastrophes. So, those friends who warn us against catastrophism are looking in the wrong direction and barking up the wrong (threatened) trees.

The three most acute catastrophic possibilities of today are nuclear war, climate wreckage, and fascism. And they are intercoded: climate catastrophe generates climate refugees flowing from south to north, secretes potential fascist movements in old capitalist states, and makes some of those states more willing to risk nuclear war. The era of climate refugees, now primed to expand radically, explodes together old images of a world of nation states and planetary gradualism.

Today *catastrophism* must be challenged by a measured, *distributed catastropheism, that is, acknowledgment that several deleterious effects are baked in and that more severe cascading disasters unevenly distributed to different regions and classes are in store unless extreme corrective actions are taken rapidly.* Measured catastropheism exposes and fends off false flag catastrophe stories, partly to expose how such mobilizations conceal real dangers. In fact, one reason that many in old capitalist states are so susceptible to false catastrophism today is that they sense that something *is* fundamentally wrong but fail or refuse to locate its sources correctly. Why? In part because many of these constituencies do suffer under existing institutions and worry that they will suffer even more if regional mitigation efforts are seriously invoked. And because many superrich plurocrats realize that acknowledgment of the real dangers would add critical pressure against modes of neoliberal capitalism and even linear images of time to which they cling. The superrich, of course, represent the real entitlement crisis of today. Hence the denialism propagated by those among the superrich, who give every sign of knowing better, and then accepted by many working-class whites clinging to white superiority. It is now clear that many of the superrich no longer believe their own denials of climate change. They are, rather, holding out and on as long as possible, planning exquisite escapes while the rest of the world remains stuck in the wreckages they leave behind. But perhaps their best-laid plans will not work either, partly because the new forms of climate wreckage may trigger wars that encompass them too and partly because the recoil effects of accelerating climate wreckage are apt to travel to their retreats as well.

We inhabit a time of tragic possibility reflected in potential, asymmetrically distributed catastrophes of the time.

In clarifying how legacies of white settler societies, the consolidation of old capitalist states in temperate zones, and the emergence of newer capitalist states in nontemperate zones combine with impersonal planetary distributors to carry the worst effects of climate wreckage to several of the world's poorest regions, to help foment civil wars, to instigate massive migrations heading north, and to incite opportunistic fascist movements in old capitalist states whose leaders obfuscate the sources of the migration drives they despise *is to fold the vicissitudes of climate heating into the heart of social theory itself.* Fascism, alongside famine and nuclear war, poses great dangers flowing from intersections between uneven climate wreckage, enhanced refugee flows, and dangerous temperate, capitalist state responses to both. For example, Venezuela, previously unimportant as a source of refugees to the United States, now spawns millions of refugees who brave the Darien Gap. They are *driven* by intercoded political insecurities, climate duress, and economic peril; they are *pulled* by promises of jobs and security in America. Such a push/pull combination regularly fosters such movements, including the drives of Mongols into Europe during the Medieval Warming Period, the rush of white settlers west from the seventeenth to nineteenth centuries in America, and the hardened attitudes of many white workers and multibillionaires today.

As the climate and refugee effects of planetary circuits now recoil increasingly back upon old capitalist states themselves, it is imperative to counter both climate denialism and climate casualism, doing so while insisting that the growing migrations of nontemperate zone peoples into temperate zone states must be absorbed and accepted. Otherwise the old states will increasingly resist "surprising invasions" they themselves help to foment, thereby playing into the hands of the fascist right. It is very clear now in the United States, at least, that the radical right intends to foment the very "border crises" they decry.

Climate casualism today is a form of climate disavowal: you acknowledge the phenomenon formally while quickly moving on to other issues often falsely treated to be separable from it. Lurking in that double perspective, I suspect, is a dark sense that capitalism and modernity cannot continue down the trails of progress upon which their institutions depend, joined to the conclusion—finding different expression in different classes—that it is too painful or too uncomfortable to do what it takes to reverse those priorities. Hence casualism.

As this book was about to go to press, a new, impressive study came out by the climatologist Michael Mann, *Our Fragile Moment: How Lessons from Earth's Past Can Help Us Survive the Climate Crisis.*[14] The richness of this book resides, among other things, in its detailed efforts to show how previous severe warming events are unlikely to provide automatic prologues to the future today. There are key differences, for instance, that make the PETM (Paleocene-Eocene thermal maximum), 55 million years ago, and the Eemian period, 130,000 years ago, unlikely to predict the same high levels of heating and ocean rise today. And he finds reasons to place a collapse of the West Antarctic

Glacier on a slower rather than faster timescale (though others even more recently say it is moving fast).

These accounts are subtle and impressive. There are things about Mann's book that concern me, however. First, he continues to use the generic "we" in discussing climate emissions and political strategy. Second, he limits his strategic engagements to governmental policy, ignoring the vital importance of social movements in forcing reluctant states, corporations, and international agencies to act. Third, and perhaps most important, the narrative structure of the text focuses on how he deviates from "doomsday" theorists such as Guy McPherson and Jem Bendell who say that extinction is locked in, as well as "breathless" news reports that receive them as gospel. Such doomsdayers must indeed be rebutted. The risk he runs, though, one finding expression in his first interview on MSNBC, is that the prime contrast-model of doomsdayers diverts attention from even more important impediments to resolute climate policy posed by deniers and casualists; and it plays down the extent to which his own account tacitly moves rather close to the measured catastropheism summarized above. Indeed, each vivid critique of a doomsday finding is followed by warnings composed in a lower key about real dangers and uncertainties accompanying accumulating trends of heating, glacier melts, ocean acidification, greenhouse releases, expected sea level rises, new wildfires, storms, and droughts. While there is a link between doomsday seers and climate denialism—both undermine climate activism—Mann plays down too much the dangers of obdurate climate casualism. Relatedly, he fails to explore the devastating effects new climate migrations and temperate state fascist responses to them may set in motion. Fascists, among many other transgressions, thumb their noses at climate wreckage as they intensify reckless capitalist modes of exploitation, repression, and fossil fuel use. Doomsdayers are important agents to criticize; fascist nihilists and climate casualists even more so.

We will turn soon to the vexing question of political strategies in the current situation. But first it is essential to turn to another feature of this historic conjuncture.

Beyond Linear, Singular Images of Time

A linear image of time is one in which time is said to follow a straight, unbroken line; a singular image is one in which time is said to be governed (or measured) by one trajectory. A linear, singular image of lived time, as Augustine and Kant saw, meshes well with evangelical versions of Christianity, broadly defined, that demand entire societies be built around faith in an afterlife and a second coming. Noble versions of Christianity that resist such universal demands today on behalf of existential pluralism may sense that they also need to reconsider the very notion of time previously absorbed. Such a unilinear image meshes well, too, with neoliberal, capitalist images of progress conveyed by a constantly growing economy. Capitalism itself is staked on such an image, as becomes abundantly clear in the diverse capitalist theories offered by Hayek and Keynes. If popular faith in the future capitalism demands were to collapse, effective capitalist planning into the singular, extrapolated future it projects would fall into crisis. If capitalism continues on its current trails, more climate catastrophes are in store.

Today, however, in the face of multiple experiences of growing climate wreckage, more and more people and institutions *cling* to a linear, singular image rather than merely inhaling it with the air they breathe. Indeed, one collective anxiety of today is about time itself, an anxiety that impels some to intensify cultural war in favor of an official image that is frayed around the edges and soft in the middle.

Long western campaigns to impugn pagan, cyclical images of time, noted earlier in this study, were stalked so fervently in part because the stalkers staked their own spiritual lives on some variant of a singular, linear image. That existential demand is particularly apparent in Augustine; it also finds expression in Kantian anxieties, in neoliberal patterns of investment and economic insistence, in evangelical patterns of insistence, and in the tendency of neoliberalism to morph into fascist versions of capitalism as the future projected becomes even harder to sustain. A whole series of economic and creedal commitments today are staked on old images of time to which some cling and which others pretend can be sustained as they quietly plan private escapes.

But what if, under the pressures of a new age, a third image began to gain more plausibility, as diverse thinkers such as Ghosh, Nietzsche, and Michel Serres suspect it should? The idea would not be to insist that it provide the universal center of society as such. But to defend it as one credible possibility, to be included within the corpus of a pluralist exploration and debate. As, for instance, Catholicism eventually became incorporated into Protestant regimes and Judaism and secularism did thereafter in the same regimes. Even if precariously. It is such an alternative that I seek to support.

We will proceed in three stages, underlining the emergence of deep time in geology, experimenting with the idea of time as a bumpy composition of multiple temporalities, and reviewing several shifts in orientation needed if you were to embrace both of those themes. To rethink climate wreckage, with its multiple, intersecting trajectories, is to rework western images of time.

Deep time. As Stephen Gould and others have reviewed, the bishop of Ireland in 1654 estimated the inception of the earth itself at 4004 BCE.

Augustine staked his theology on that assumption and Newton accepted it too. The great naturalist Buffon in 1779 estimated the age at 75,000 years. In 1862, the physicist Lord Kelvin, using estimates of the rate of earth cooling, tried out 100 million years. Darwin is said to have guessed an age of 300 million years because that was how long it would take, on a planet with a roughly uniform geology persisting through time, for species evolution to reach its high complexity. After the rate of radioactive decay was discovered and rock dating became more sophisticated, Arthur Holmes estimated the age to be 1.6 billion years. By 1921 it was estimated to be 4.53 billion years old, give or take 50 million years on either side. Today, the best estimate by geologists is roughly 4.55 billion years. Western time increasingly sinks into deep time.

Several of these dating episodes created a flurry of debates when they were advanced, some between physicists and geologists, the most dramatic between devotees of the biblical Genesis historically embraced by church fathers and the sciences. We will bypass those debates here and now, even though they spawned real effects. Even though, also, hesitancy among geologists for two centuries to accept the idea of bumpy, geological time may well have been rooted in the quest to avoid new intensification of cultural war with evangelicals.[15] Georges Cuvier, the great French planetary catastrophist, for instance, was accused by geological opponents such as Charles Lyell of opening the door once again to a theology of end times.

A philosophy of deep time—let us mark its struggle for precarious acceptance in Euro-American cultures from the time of Darwin to today—provides ample room to chart planetary movements and species evolution. But it harbors ambiguities. Deep time can, for instance, be set in an assumption of slow, planetary gradualism, as Darwin and Lyell did along with most of their Anglo-American earth science followers for two centuries, even while some in France and Germany held out against it. These two, I suspect, absorbed remnants of a Christian image of time, even after calling into question the god who authorized it.

To both accept deep time and overcome planetary gradualism, Michel Serres recently began to experiment with the idea of time as *multiplicity*, an image that both requires and amends previous images of deep time. It opens up, you might say, cyclical images, appreciating the multiplicity of interacting cycles and promoting the idea that new formations can be organized out of such intersections. It also exceeds linear and singular images by treating time as a multiplicity. And it treats clock time as a useful dating of events, but denies it the honor of representing time itself, even though Descartes, Newton, and Einstein, in different ways, had done so. Each of these three images is useful at times; all three, on this new reading, are absorbed into time as multiplicity.

Serres—a thinker indebted to European pagans such as Lucretius and Archimedes as well as to new turns in Euro-American science after the advent of quantum theory—contends that time consists neither *solely* of the uniform, singular flow represented by Descartes nor the singular repetition of instants curving under the influence of gravity represented by Einstein. Einstein, to him, is too dead set on treating a single measure of time—the speed of light—as *the* measure. To me, Einstein may have thought he had disconnected time from experience, but in fact he has defined one experience—the speed of light—to be sufficient and then clothed that experience within a binary logic.[16] Serres would find the Bergsonian image of duration appealing, appreciating as it does how experiences of past, present, and future flow into one another. But it is insufficient. Philosophers of durational time—James, Bergson, and Whitehead leaders among them—may not appreciate the plurality and bumpiness of such flows sufficiently. Some may not fold the larger plurality of species experiences they themselves project sufficiently into the periodic bumpiness and plurality of nonliving planetary forces.

All such images are too streamlined for Serres, too governed by a single problematic metaphor or measure, too invested with eagerness either to commune with a benign world or to master a unilinear one. Break, break, my friends, the obdurate hegemony of the debate between such images. Things may be extracted from each. But as time bumps along, the hegemony of that debate is killing us.

We might, Serres says, start by appreciating how time *percolates* more than it *flows*. That image turns you first to thinking about the turbulence of boiling water. Serres extends it to the turbulence of streams and rivers. When I first encountered that metaphor, I was drawn back to the stick races I used to stage with my children on this or that bubbling creek. One well-designed stick would take an initial lead and then whirl around in an eddy. Then it might turn and be blocked by a pile of rocks and branches before, if the youngster rooting on the bank were lucky, it pulled out. It might even be pulled by a subcurrent upstream for a while; or sink into a whirlpool only to pop up somewhere else. The hypersensitivity of the stream to minor perturbations is pertinent here.

The winner of the race is not always the one with the stick best fitted to flow in a stable current by weight, length, volume, and density, though those features induce advantages. Other contingencies enter the fray. Time percolates, even *if* you start with a river or a stream as the privileged image through which to think time. Suppose, too, that downstream there is a waterfall, pulling the flow at an increasingly rapid rate until it pours over a precipice. Such an image suggests that time may speed up and slow down.

Yes, from the vantage point of clock time—tick tock, tick tock—some percolations last longer than others; but their courses are often neither uniform nor regular. Along the way a porous stick may absorb new elements, and fungi and bacteria will eventually seep into it. It might lose the race and acquire roots, or fertilize bushes on the riverbank. The "winning stick," on the other hand, might remain a mere stick longer, as measured by the clock. Here are a few other things Serres says about time:

> Sea, forest, rumor, noise, society, life, works and days, all common multiples. . . . I'm trying to think the multiple as such, to let it waft along without arresting it through unity . . . I am now attempting to rethink time as a pure multiplicity.[17]
>
> . . . like the percolating basin of a glacial river, unceasingly changing its bed and showing an admirable network of forks, some of which freeze or silt up, while others open up.[18]
>
> I am disquieted; therefore I exist.[19]
>
> . . . the entirely subjective belief in a globally positive or, on the contrary, globally negative gradient of a simple and linear curve seems naïve today; global progress or general regression, this depends on your digestion . . . [20]
>
> Time flows in an extraordinarily complex, unexpected, complicated way . . . it folds or twists; it is as various as the dance of flames in a brazier—here interrupted, there vertical, mobile, and unexpected.[21]
>
> For whatever praise you may hear, whatever love you may profess for the sea and mountains, the desert or marshes, plants and animals, nature doesn't behave as a friend to humans or even as their symbiont. By means of waves, fire, typhoons, poisoning or devouring, it kills as calmly as bodies fall and eagles eat lambs.[22]

Through such formulations, Serres seeks to break the existential hold old images have over binary logic in general and the primacy of mastery/organic debates about time in particular. The old western *logics* depress attention to messy processes of becoming that exceed them. The image of time divisible into instants pretends the world is mostly composed of solids and neatly divisible into parts. It underplays the fluidities, crossings, pluripotentialities, and becomings that also populate the earth.

Construing time to consist of intersections between multiple temporalities, Serres teaches us to master the will to mastery over the earth.[23] Doing so, first, because it cannot succeed in a world of multiple, intersecting temporalities and, second, because it fails to respond affirmatively to the veritable grandeur of that of which we ourselves form a minor part. The aggressive nihilism with which many respond to the last sentences quoted above from Serres is, to Serres, a sign that they have not yet overcome the profound disappointment that their favored images of time do not mesh well with actual encounters they have had. It is that *cosmic* disappointment we are ethically enjoined by him to overcome, so that our thinking and responses to the world become more decent, in-formed, and imbued with presumptive generosity. For existential disappointment, unless overcome, readily morphs into existential resentment. *And resentment against the character of the cosmos* can readily evolve into angry constituency dispositions to destruction of human scapegoats.

I will try to keep these suggestions and admonitions in mind as I draw from two of my mentors—Gould and Serres—to think further about relations time may bear to multiple temporalities of heterogeneous sorts.

What follows, then, is an effort to absorb recent geological evidence about radical turns of time in the deep past into the Serres's schema of planetary time as multiplicity.[24] I construe that to mean that time is always tied to experience. But the experiences are not confined to humans alone. They include those of living beings of diverse sorts *and* the periodic sensitivities of nonliving forces. Time is composed of multiple temporalities moving at different speeds and trajectories that periodically intersect. So it may be wise to think of *timescapes*—periods (in clock time) when multiple temporalities of diverse sorts intersect, bringing new events into the world.

Consider, to bolster such a theme, several dramatic planetary and global events, some of which have been introduced briefly for other reasons earlier. Their accumulation now may help us think time as a bumpy multiplicity replete with timescapes composed by intersections between diverse temporalities.[25]

Perhaps between 4.1 and 4.5 billion years ago a planet, now called Theia by geologists, probably crashed onto the earth, which then was still covered with molten lava seas. A collision of two cosmic temporalities big enough to make the Greek gods stutter—one planet on its trajectory, another on a different trajectory. The result seems to have been the formation of a moon around the earth, with effects (later) on the tides of the earth. Some geologists also believe that Theia affected the future tilt of the earth and deposited "carbonaceous material" on earth from which oceans were later formed. The jury is out on those two theories, but, if true, they combined to set new temporal conditions for seasonal life on earth. Billions of years later, sailors deployed the light of the moon and its orbital course to help guide them at sea. Adventurous migrations to numerous Pacific islands were enabled by that collision. Another event in a veritable timescape of colliding temporalities.[26]

About 3.7 billion years ago life emerged and began to evolve, at first in the oceans. About 360 million years ago some plants began to move onto land, enabled by the oxygen revoluton promoted by photosynthesis. Later, land plants evolved to form lignin, a hard material allowing trees to grow high. For about 90 million years, dead trees would pile up, slowly forming carbon materials that would much later provide coal and oil to humans. The emergence of fungi, with impressive lignin-eating capacities, halted that process, meaning that the clock window for formation of oil and coal was only about 90 million years. If, say, these fungi capacities had evolved millions of years earlier, the modern rise of climate wreckage might not have taken the turns it has.[27]

About 250 million years ago (in deep clock time), massive eruptions from the Siberian flats heated the earth's atmosphere, because sun rays could penetrate the cloud cover, but many particles reflected back into it were trapped in the atmosphere (three events). Then (again in clock time), the warming atmosphere probably released massive methane sediments in Antarctica (another event), heating the planet so much that 90 percent of life was lost. If the methane release had been larger yet, life might have been destroyed totally, waiting millions of years to be activated again, following different trajectories than those actually taken.[28]

The massive asteroid that hit the earth 66 million years ago is now familiar to many. It is only one of many such turning temporalities, however. Moreover, the asteroid itself might have been formed slowly through the accumulation of ice, dust, and gravity. And then, after that massive impact, a several-year break in photosynthesis, created by its release of a huge volume of airborne particles, clinched the incredible loss of species life. Millions of years after that event, its belated projection into the past by Luis and Walter Alvarez in 1980 also fomented a huge conflict in the earth sciences that had been organized around the assumption of planetary gradualism. Another bump in *scientific* temporality. If this new projection were true, then gradualism faced revision, and future advances in the earth sciences would require extensive cross-disciplinary work between sciences that had heretofore sought to remain rather autonomous. After a decade of intense struggle the Alvarez projection was supported by enough evidence and argument to become secure. The crater itself, for instance, was not isolated until 1991. The accumulating evidence in favor of this asteroid explosion thus helped to foment veritable revolutions in geology, evolutionary theory, climatology, paleontology and, much more belatedly, philosophy and social theory, as each field now felt pressure to become more closely imbricated with the others. Did it eventually help to set the stage for explorations of time as a bumpy multiplicity that itself helps to jolt images of earth mastery?

About 12,700 years ago, the worldwide ocean conveyor system, which had followed a *cyclical pattern* for a few million years, suddenly stopped. The Gulf Stream, part of the conveyor, halted. The impetus for the stoppage is still under investigation and the stoppage itself was not discovered until the 1970s in clock time. But its effects on human and nonhuman life were enormous. A thousand years or so later, the current was renewed and the Holocene stuttered into being, perhaps over a period of less than ten years. Another turn of time . . . [29]

About 7,000 years ago, the rapid rise of the Mediterranean seems to have opened the Bosporus straits and generated the Black Sea in the basin where a smaller freshwater lake had been. Human survivors of that horrendous event may have authorized the sacralized story of Noah and the Ark, as well as other flood stories carried forward by Mesopotamians and Ovid.[30] A fateful, turning event composed of multiple temporalities.

About 2,000 years ago, Jesus was born of an unwed mother in an outpost of the Roman Empire. His ministry eventually spread to several continents, helped to spawn a new European church, entrenched a European image of time, and eventually lent support, though ambiguously, to invasion by Europeans of the American continents.

By 1619 Portuguese sailors kidnapped Africans from Kongo and Ndonga, terrorizing their communities and shipping survivors to Jamestown, Virginia, helping to inaugurate hundreds of years of brutal chattel slavery in America to build a "new world" on the backs of enslavement.

About 125 years ago (1897–99), seasonal monsoons (a cyclical temporality) were interrupted over large parts of India. The interruption seemed to follow an intensification of El Niños over the Pacific and a shift in the direction, intensity, and absorptive capacity of western trade winds. The British Empire failed to respond to the massive famines and disease that followed. Five temporal conjunctions, then, between an intense El Niño formation, shifting wind currents, seasonal monsoon interruptions, and the policies of the British Empire. To manufacture devastating famine.[31] Creating a new timescape.

This mere sampling of highly *diverse* events—and the multiple conjunctions between heterogeneous temporalities that fomented such timescapes—is incomplete and arbitrary. More instances could be added, the more multiple and heterogeneous the temporalities involved, the better. Even this limited sample, however, may help to mark time as a bumpy multiplicity grounded in disparate experiences, sensitivities, and temporal intersections. It is not that each event lacked preconditions, but each arose through new intersections and often came as a shock to those affected. Sometimes an event would build slowly and then shift much faster upon activation of a tipping point.

Such events emerge from the confluence of two or more temporalities, previously set on different vectors, speeds, and capacities. The confluence *tacks, turns, twists, accelerates, or decelerates* the temporalities preceding it, now making any linear "extrapolation" based on preceding trajectories—to use Darwin's favorite and overused word—out of touch with the turn or acceleration actually taken.

Refusals to adjust linear extrapolations into the future in the face of new events may provide one impetus to upsurges of fundamentalism, white supremacy, climate denialism, casualism, and fascist nihilism in Euro-American societies as climate wreckage becomes more volatile. Some of those disparate refusals may involve desperate desires to save an old faith in a second coming, or to preserve a reassuring image of progressive time, or to pretend that whiteness is superior, or to protect the assumption that the linear growth of capitalism can proceed indefinitely. Or several of these in alliance. Hence the counter call to develop a philosophy of time as a multiplicity composed of heterogeneous temporalities.

To give events their due, physicists, theologians, humanists, and social theorists must become rather more familiar with bacteria temporalities, viral temporalities, fungal temporalities, civilizational temporalities, capitalist temporalities, glacier temporalities, climate temporalities, geological temporalities, colonial temporalities, asteroid temporalities, scientific temporalities, ocean current temporalities, and theological temporalities, noting how any of these may intersect with one or more of the others at key moments. Thereby spawning a new event. Time, multispecies experiences, diverse planetary temporalities, and the advent of new events now become intercoded.[32]

To think time as multiplicity during the latest time of climate wreckage does not eliminate the need or ability to address continuities. The tilt of the earth has persisted for billions of years in clock time, once it jolted into place. And even though it is undergoing minor shifts today because of massive glacial melts and reductions in the level of groundwater aquifers. An ocean current may last millions of years. A capitalist regime for hundreds. But an event, often composed through an intersection between two or more temporalities, bends, twists, accelerates, or decelerates previous continuities. An event might start with a bang or begin with a whimper and then accelerate. Time as multiplicity disorients or disquiets partly because it unsettles assumptions and temporal extrapolations deeply engrained in western cultures. Sometimes events overwhelm.

A Few Implications of Time as Multiplicity

Here are a few lessons brought into relief by a focus on time as bumpy multiplicity. First, previous major events—previous turns of time—continue to carry implications today (in clock time). The shape of seasons, the current hegemony of *Homo sapiens*, the flow of the ocean conveyor, seasonal monsoons, the current conditions of indigenous peoples, the aftereffects of racialized enslavement, the current trajectory and pace of climate change, the increased danger of nuclear war, all find expression in part because of events that turned, dampened, or intensified previous zones of time. Previous turns fold into current patterns.

Second, when an event erupts, old species extrapolations into the future may now need to be adjusted in small or large ways. The recurrence of events *suggests that more of us should cultivate a double orientation to temporal extrapolations*. During periods of reasonable regularity in a domain—such as regular seasons, a settled civilizational pattern of class, gender and race hierarchy, the confinement of viral crossings, a slow pace of species evolution, the consistency of climate patterns, the reliability of monsoons, the relative stability of democracy, etc., etc.—it is often reasonable to extrapolate forward probabilities from the recent past to the future, adopting supportive or critical social movements by reference to those extrapolations of probability.

During more volatile periods, including the advent of the Anthropocene, new extrapolations of possibility, probability, and desirability are needed. Yet today too many people not only cling to interests and faiths that discourage such revised extrapolations; they also cling to an image of time that does so. A progressive, linear image of historical time, for instance, meshes well with planetary processes said to change on a long, slow trajectory.

One source of fascist energies today resides in the fact that fewer people in old capitalist states either believe in their bones in the benign future of capitalism *or* in the most familiar productivist alternatives traditionally ranged against it. All of them tend to be set in linear images of time. Indeed, the refusal to alter old extrapolations when a specific zone of time turns or accelerates helps to foment patterns of denialism and/or scapegoating in many circles today.

Those who acknowledge time as a multiplicity are thus encouraged to adopt a *double-entry orientation* to temporal extrapolation. Extrapolations into the future are always needed, but when big turns or accelerations occur, new extrapolations of future danger and possibility become pertinent. That is, a double-entry orientation emphasizes the imperative to revise embedded extrapolations when a significant turn in time occurs.

Third, the recurrent intrusion of bumpy events into life—that is, of emergence—also means that entire cultures are periodically called upon to fend off the existential disappointment, or even cultural rage, that readily arises when old debates about time—debates, say, between organic and progressive images and between cyclical and linear images—*no longer control the cultural terrain with assurance.* The challenge now becomes to encourage more constituencies to appreciate the grandeur of bumpy time, to affirm a world of multiple temporalities curtailing old confidences in probable progress along a unilateral track. Appreciating the grandeur of bumpy time to encourage struggles against the worst things when a bad turn, deceleration, or acceleration occurs. Doing so without seeking racial, religious, scientific, and theological scapegoats. For temporal trajectories periodically do turn.

Recent experiences with the volatility of the world—including the advance of droughts, new threats to monsoons, the rapid spread of viral crossings, extrapolated increases in volcanic eruptions as the climate warms, rising seas, the enhanced staying power of storms, the intensification of human migration drives, volatile responses to such migrations in old temperate zone, capitalist states, and so on—mean that the issue of time as multiplicity cannot be hustled offstage without severe consequences.

One dilemma of today is that too many constituencies and institutions—including capitalism, Christianity, and democracy—remain staked to images of time that may no longer be sustainable. They are also staked to a human exceptionalism that pays too little heed to multispecies vitalities and multiple planetary temporalities.

Marx has been widely held to be a theorist tied to a socialist version of the exceptionalist, mastery, abundance club. Including by me, until recently. I had at first been attracted as a young man from a working-class family to Marx as a theorist of alienation, then moved away as he became more closely identified with Althusserian modes of mastery in the American academy. I now find myself seeking to draw upon him again as new work on his notebooks after 1868 appears. A recent book by Kohei Saito, entitled *Karl Marx's Ecosocialism,* severely qualifies the image of Marx as a promethean socialist. Marx advanced the former themes up through volume 1 of *Capital,* says Saito, but previously unpublished notes composed after 1868 show that he became very concerned about the effects of deforestation, began to read scientists who studied metabolic relations between capitalism and nature, and even concluded that human-influenced climate change was an issue to address.[33] The Saito study is impressive. As these notebooks continue, Marx becomes less confident that the advance of capitalism will pave the way for a new communist society of productivity, more concerned that its historical trajectory promotes collapse. As the author writes, the manuscripts in question are still being edited and examined.

Much more needs to explored about this shift. But one set of adjustments can already be proposed. Marx, of course, was profoundly insightful in showing how capitalism renders workers insecure and alienated from their own work process and the products they produce. Today, as more people come to appreciate capital/nature relations, and even time as multiplicity, three additional modes of alienation might also be brought front and center. Alienation from death, alienation from the shaky place of the human estate on a periodically tumultuous planet, and alienation from a world populated by numerous nonhuman agencies. These latter modes of alienation, however, do not need so much to be *overcome* as to be *transfigured.* To the extent possible, more of us must strive to accept death in old age as a condition of life itself. More must also seek to fold into relational practices appreciation of how bumpy multiplicities of time create asymmetrical dangers for humans and other species as they also provide the sublime seedbed from which life itself has emerged. And more of us must strive to become finely attuned to the multiplicity of life-forms and strivings that populate the earth. Even as we struggle to respond to the death flight of capitalism.

An Improbable Necessity

In a 2017 book entitled *Facing the Planetary* I joined appreciation of the urgency of time to ethological work upon the swarming practices of honeybees to articulate an "improbable necessity," a multiform, cross-regional "politics of swarming" both historically improbable and historically necessary. Necessary because without its different dimensions working back and forth upon each other, climate wreckage would overtake, if differentially, every place on the globe.

Such a politics invests in several dimensions. The first involves an escalation of *role experiments* to alter behavior with respect to the consumption you practice, the recycling you do, the teaching you undertake, the speakers you invite to your church, the clubs you join, the blogs in which you participate, the unfamiliar inspirations you solicit, the co-ops you form, the dream life you solicit, the things you say at meetings and talks, and so on. Pursuing such role experiments promotes cumulative reductions in emissions *and,* even more, primes us to participate in larger collective efforts. Not every individual and constituency in a class society can do all these things, certainly. But many can adjust this or that variant to their own situation.

As the first practices sink into the visceral register of constituency life, they may help more people to appreciate the grandeur of a rocky planet that has enabled us and other species to emerge and to be. The role experiments, if and as they become persistent, must also be punctuated by periods of dwelling in which you experience more vibrantly bumpy relations between elements of wildness in the world and in your/our affect-imbued thinking. As when, to return to an earlier example, you sit again on the edge of a bubbling brook, allowing its arrhythmic sounds and exquisite sensitivities to perturbation to infiltrate your memory, feelings, and responsiveness. Resonances back and forth between those percolations and an element of wildness in thinking can sometimes help to foster more sensitive modes of responsiveness to the world. In relation to an issue you have been pondering, a new thought may now bubble up. For thinking is akin to a bubbling brook, and the two can be encouraged to resonate together. Through such experiences, attachment to a bumpy earth may now slip more densely into the sediments and prompts of being, an attachment reaching deeper into the well of being than encouraged by the subject/object duality still inscribed in western languages and institutional habits. Attunement to a world of bumpy emergences and birthings, as well as to some relatively persistently modes of being. Earthy gymnastics.

For the politics of swarming to get off the ground, it can thus be invaluable to dwell periodically in sublime attention to a bubbling brook, rhythmic ocean waves, jaggy mountain vistas, wild volcanic flows, the strange dynamism of a glacier melt, the visceral memory of a tornado, a fateful birthing event, the strange surge of a new love, appreciation of the wondrous evolutionary achievement of the platypus, a burst of laughter from nowhere as you sense something ridiculous in what was just said or done, close attention to dynamic resonances by which a virion crosses from one species to another. The first experiences solicit modes of attentive responsiveness. These explorations might even help—in tandem with corollary changes in theory—to enhance attachment to the earth, folding a more robust appreciation of its grandeur and volatilities into us as we loosen within ourselves and our constituencies some of the closures in identity and constituency life that so readily become attached to both. So, too, does attentiveness to previous struggles, as when you attend to critical Amerindian movements today and in the past. The eco-alliance between the Lakota, ranchers, and urban ecologists a decade ago, for instance, can help to stimulate new assemblages today.[34] Together, role experiments, dwelling in sublime experiences, and study of relevant social movements may weave appreciation of the grandeur of a bumpy earth more finely into our souls, our thinking, our identities, our communications, and our political strategies.[35]

As the role experiments and these practices of heightened sensitivity work back and forth upon each other, we—the we here is invitational—may well feel more compelled to join escalating demonstrations and electoral politics to respond more militantly to climate wreckage, doing so to force a wider spectrum of radical, rapid moves. These latter events, too, can, if accompanied by luck, both promote positive political effects and inscribe respect for the earth more deeply into our souls.[36]

Let us laugh together on principle about previous bouts of anthropocentrism and sociocentrism, as we mobilize politically around the serious issues of climate wreckage.

As those three practices increasingly reverberate, it may soon become timely to instigate cross-regional climate strikes, focused on established modes of production, consumption, state investment practices, and interstate priorities. Doing so to place internal and external pressure upon capitalist states to reshape radically their economies and to reconstitute the cultural ethos residing in economic practices. For such changes will involve massive and experimental undertakings during the new time of climate wreckage. The politics of swarming encourages these several dimensions to work upon each other, until each becomes much more than it would have been alone.

There have been ripples of such efforts here and there, with growing numbers of young people in multiple regions seeking to spark cross-regional strikes. But the needed sense of urgency is still not woven sufficiently into several constituencies. Some insistently make old extrapolations into the future with greater belligerence; others adjust their thinking on the refined registers of political discourse while retaining remnants of previous images of planetary gradualism, linear time, and the politics of growth on the cruder, visceral registers of cultural action and communication. I call the latter modes of disavowal "passive nihilism," finding it to be a rather common cultural orientation among disoriented constituencies in old capitalist states today, an orientation ill equipped to respond either to climate wreckage or to the growing belligerency and callousness of the fascist right.

Today—this text is being completed in the winter of 2023—new extrapolations into the probable, short-term future are needed. They translate what was previously a sense of urgency into one of near desperation. Things continue to get worse with respect to climate wreckage, and, as we have seen, its amplifications both promise to magnify urgent refugee migrations from nontemperate into temperate states and threaten to excite new fascist energies of reaction in those old capitalist states. Those, today, pose extrapolations to make of danger on this bumpy planet.

A politics of improbable necessity resists such probabilities, even as it supports massive reparation and mitigation projects. Today, more of us need to express greater urgency yet about the general dangers and uneven distributions of climate wreckage as we participate in the politics of swarming. Both.

The task is not so much to activate the conscience of the superrich in old capitalist states, though there are a few exceptions. The extreme entitlements they now receive and demand—as they control a monstrous and ever-growing portion of social wealth and contribute much more than their per capita share to climate destruction—place most beyond reach. They too often fashion private escapes, finance venture capitalism, and support or tolerate authoritarian movements. The task is to dramatize first and foremost to young people in diverse social positions and regions multiple species vitalities, to accentuate communications between minor European traditions inspired by pagan history, indigenous peoples, and eccentric Christianities, and to become more militant in staging radical actions and strikes against old capitalist states and newer members of the high-extraction club. Such movements also invite participation by other constituencies, doing so because it is now very clear that white power movements, Christian fundamentalism, anti–climate action, demands for heterosexual exclusivity, and fascist drives intersect.

Such countermovements are both improbable and even more urgent than heretofore. They are also possible. Perhaps as more activists attuned to a bumpy world press such energies forward, a new event or two will dramatize the dangers and activate constituencies now on the sidelines or cusp of action. As happens when hot water enters a metastable state before beginning to boil. The casual observer would have said it was stable, but metastability sits at the threshold of a tipping point by which things soon change fast on multiple fronts.

Such is the urgent hope of today. The future of a world without the spread of yet more massive climate wreckage and intense fascist reactions to its consequences in old capitalist states depends upon such activations. The improbable necessity of a multivalent, cross-regional politics of swarming.

Acknowledgments

This book has unconsciously been in the works for a couple of decades. I have taught sections of it in graduate and undergraduate seminars and composed preludes to it in this or that essay or book. Now those preludes have been folded into a critical cosmology, speculations about two strains of paganism, and the current time of climate wreckage. This book also, however, contains limits that I have tried to surmount in other studies. The reader will find engagements here with the ethos of white settler society but not a close account of modern capitalism, in its Keynesian, neoliberal, and fascist variants. My hope is that the account offered in chapter 2 of *Climate Machines, Fascist Drives, and Truth* makes up for that deficit here. The intercoding between a capitalist *axiomatic* and variations of a capitalist *ethos* in that chapter meshes, I believe, with the accounts of two settler patterns of conquest and eco-destruction in this text.

In the fall of 2021 P. J. Brendese and I cotaught a graduate seminar that explored several issues in this study. Indeed, his new book, *Segregated Time*, is relevant to my efforts, and I am indebted to him and it. I am also inspired by the fifteen students who participated so actively and thoughtfully in that class. Their presentations on the likes of Hesiod, Ovid, Augustine, Viveiros de Castro, Tzvetan Todorov, Bill McGuire, and Amitav Ghosh were always lively and filled with insights that, I hope, help to illuminate this study. It was, you might say, a wonderful way to close a long teaching career, even if the topics were often bruising. We inhabit a bruising time and are, variously, inhabited by it. Cultivating and protecting a mood of thoughtful care for the world is indispensable under such circumstances. The participants in that seminar were acutely attuned to this triple condition.

In general, I have been extremely lucky to have had so many thoughtful graduate students over the years; they have both inspired me to keep going and added new strains of thought and aspiration to fold into my work. In that respect a Hopkins Symposium entitled "Theory in Bold Strokes" in November 2023 brought together many intellectuals who think deeply about these issues. I am indebted to the papers and questions posed at that event.

Numerous colleagues have been very helpful to the pursuit of this project. Tom Dumm and I talked about it often, and he commented invaluably on all the key chapters in their early, very rough drafts. I always profited from his thoughtful suggestions. Jairus Grove, Steven Johnston, Nidesh Lawtoo, Claire Sagan, Bradley Macdonald, Stewart Motha, Catherine Keller, Tvrtko Vrdoljak, Derek Denman, Anatoli Ignatov, Smita Rahman, Kathy Ferguson, Libby Anker, Naveeda Khan, Mike Shapiro, Chad Shomura, Lori Marso, and Stephanie Erev also helped me to get my bearings and to hold onto them once they arrived.

I recall with pleasure how the interview with Stewart Motha on the podcast *Countersign* in 2022 helped me to consolidate these themes. An exchange with Dipesh Chakrabarty at the Presidential University in India also helped to clarify these thoughts. I appreciate, too, the invitation from Lars Tønder at the University of Copenhagen to speak to a large, vibrant audience in Copenhagen on these topics. The questions he and others posed about the pros and cons of "catastrophism" both helped me to think through that issue and showed me how many young people are ready today to explore harsh themes without lapsing into cynicism or despair.

In December 2022, Heike Harting, Heather Meek, and I pursued a three-way conversation that will find its way into their edited book on planetary health and climate change. That dialogue was very helpful to me; I hope some of the insights emerging from it have found their way into this book.

Ashley Kim read these chapters in an early form and helped me considerably as I revised them. David DeBole entered the process later. His textual suggestions were also very helpful, as was his superb work on the index and bibliography. These two theory grad students are highly reflective, and I am lucky to have worked with them.

Tom Lay, my editor at Fordham, has worked closely with this project from its inception. He helped me at every stage of this study, supporting the idea of chapter/coda oscillations when a few others thought it broke up the continuity of the text. In a world of multiple, intersecting temporalities of heterogeneous types, thought needs to track both continuities and breaks. I thank Tom for his reflective contributions to this text. Nancy Basmajian, an acute copyeditor who communed with this text as she edited it, was also a valuable and appreciated ally.

Finally. I thank once again Jane Bennett. Her exploratory work continues to inform me, as it does so many others. She made valuable suggestions about this project as it proceeded. Most often, after facing a brief period of resistance, they found their way into the final version. Coteaching with Jane over the years has been wondrous. She is an existential inspiration.

Notes

Introduction: Lived Cosmologies and Climate Wreckage

1. In *The Physicist and the Philosopher* (Princeton, NJ: Princeton University Press, 2015) Jimena Canales reviews superbly the wandering century-long debate between Albert Einstein's image of time and that of Henri Bergson. The former anchored "objective" time in a binary logic and newly refined technologies. The latter accepted Einstein's account of how light rays bend under the force of gravity, but he denied that this temporality exhausts the complexities of time itself. Both theorists, it can be said, anchored time in different modes of human subjectivity: Einstein in the human production of binary logic and high technologies, Bergson in the recurrent experience of how the past folds into the present and the present into an anticipated future. I review that debate briefly in William E. Connolly, *Resounding Events: Adventures of an Academic from the Working Class* (New York: Fordham University Press, 2022), 175–81. The argument there is that the later development in numerous sciences located *between* the logic of relativity theory and everyday life, sciences such as virology, bacteriology, geologies of diverse planetary temporalities, quantum theory, ethology, and so on, means that the yawning gap between physics and popular cultural experience Einstein identified no longer suffices to cover temporal scapes. This book, in a sense, starts at that point, as it emphasizes how lived cosmologies infiltrate various responses to the world.

2. Quoted in Christopher Flavelle, "As Groundwater Dwindles, Powerful Players Block Change," *New York Times*, November 24, 2023, https://www.nytimes.com/interactive/2023/11/24/climate/groundwater-levels.html.

3. Aspects of these orientations are insinuated into the visceral register of cultural life, as we shall see. For a history of "mimesis," including how modes of affective communication and more refined conceptual communication work back and forth on each other, see Nidesh Lawtoo, *The Phantom of the Ego: Modernism and the*

Mimetic Unconscious (East Lansing: Michigan State University Press, 2013). The current study rests upon similar assumptions.

4. For some representative and diverse examples, see Barry Commoner, *The Closing Circle: Nature, Man, and Technology* (New York: Knopf, 1971). That book, organized around the proliferation of waste, pollution, and the need for sustainability, helped to shape Michael H. Best and William E. Connolly, *The Politicized Economy* (Lexington, MA: D. C. Heath, 1976 and 1983), particularly a chapter called "Nature and Its Largest Parasite." See also Jürgen Habermas, *Legitimation Crisis,* trans. Thomas McCarthy (Boston: Beacon, 1973; the German version appeared in 1970); Fred Hirsch, *Social Limits to Growth* (London: Blackwell, 1977); Charles Taylor, "Interpretation and the Sciences of Man," *Review of Metaphysics* 25, no. 1 (Fall 1971): 3–51; James Baldwin, *The Fire Next Time* (New York: Vintage Books, 1963). I read the latter book when it first came out; its linkages between antiracism and a tragic vision of possibility have stuck, as will become clear in the Second Coda. See also Mary Daly, *Beyond God the Father: Toward a Philosophy of Women's Liberation* (Boston: Beacon, 1973).

5. Nancy MacLean, in *Democracy in Chains: The Deep History of the Radical Right's Stealth Plans for America* (New York: Penguin Books, 2017), shows how the neoliberal right publicly pushed for state deregulation during this period while secretly organizing to lock in constitutional control of state and national politics. For the corollary history of how a violent white power movement preceded January 6, 2021, in the United States by several decades, see Kathleen Belew, *Bring the War Home: The White Power Movement and Paramilitary America* (Cambridge, MA: Harvard University Press, 2018). Belew shows how white racism is regularly imbricated with Christian fundamentalism, gender subordination, heterosexism, anti-Semitism, and opposition to other minorities. It is also pertinent to see how the attractions of the violent spirituality energizing this movement exceed the class and racial bases from which it starts. I thank Tom Dumm for calling this book to my attention and exploring its links to themes in this book with me. To explore a fascist spirituality or constituency ethos, along with the chronic disruptions and disciplines that help to foster its mobilization, see William E. Connolly, "Bodily Stresses, Cultural Drives, Fascist Contagions," *Theory & Event* 25, no. 3 (July 2022): 689–709. Naomi Klein, in *Doppelganger: A Trip into the Mirror World* (London: Penguin, 2023), explores, among other things, how fascist forces now hijack grievances and terms from the left, twisting them into fascist energies. She used to chastise "postmodernists" for talking about how discourse has become detached from the energies of activism. Now she advances subtle and frightening examples of how new fascist movements, adopting and distorting key terms from the left, do so. The result is a series of hijacks that cloud discourse and threaten to freeze the left. I concur with her account.

6. Quoted in Peter Frankopan, *The Earth Transformed: An Untold Story* (New York: Alfred A. Knopf, 2023). This book is a veritable encyclopedia of such transformative events, often linked to other cultural processes that either exacerbated or ameliorated them.

7. Several of these examples are taken, almost at random, from Hubert Lamb, *Climate: Present, Past and Future*, vol. 2, *Climatic History and the Future* (New York: Routledge, 1977), 3–16. A recent geological study that places the Noah event in the rapid formation of the Black Sea 7,000 years ago is William Ryan and Walter Pitman, *Noah's Flood: The New Scientific Discoveries about the Event That Changed History* (New York: Simon and Schuster, 1998). The linkage between that event and the Noah story is contested by other sources.

8. The article is reported in David Lipsky, *The Parrot and the Igloo: Climate and the Science of Denial* (New York: W. W. Norton, 2023), 54. This book reviews a host of climate scientists before 1970 who addressed the issue of climate heating below the sustained attention of other scientists and larger publics.

9. Lamb, *Climate*, 22.

10. The story of how planetary gradualism prevailed in the Euro-American earth sciences for two centuries, and how it was finally subdued (at least in the earth sciences) during the long decade of the 1980s, is superbly told in Michael Benton, *When Life Nearly Died: The Greatest Mass Extinction of All Time* (London: Thames Hudson, 2005). He also focuses on the largest mass extinction 250 million years ago, venturing a theory about how the earth could have become so incredibly hot during that period in the absence of human intervention. That account also helps us understand how capitalist CO_2 and methane emissions of today and yesterday trigger planetary amplifiers and distributors that render the results much larger than the initial triggers and often distribute them to nontemperate regions far from their sites of origin.

11. For a review of his findings, and the controversies they spawned, see Wally Broecker, *The Great Ocean Conveyor: Discovering the Trigger for Abrupt Climate Change* (Princeton, NJ: Princeton University Press, 2010).

12. I do not mean, of course, that no studies have probed such wide connections. One fine example is Jairus Victor Grove, *Savage Ecology: War and Geopolitics at the End of the World* (Durham, NC: Duke University Press, 2019). Sharon R. Krause, in *Eco-emancipation: An Earthly Politics of Freedom* (Princeton, NJ: Princeton University Press, 2023), pursues a variety of strategies to respond to climate wreckage. There are numerous others. Grove, among other things, explores how academic enclosures (departments, fields) too often let climate acceleration and the effects it promotes slip through cracks between them.

13. For an array of recent examples of such emergent asymmetries, see Adam Welz, *The End of Eden: Wild Nature in the Age of Climate Breakdown* (New York: Bloomsbury, 2023).

14. These forces were earlier explored in Connolly, *Capitalism and Christianity, American Style* (Durham, NC: Duke University Press, 2008).

15. Pekka Hämäläninen, *Indigenous Continent: The Epic Contest for North America* (New York: W. W. Norton, 2022). See also Ned Blackhawk, *The Rediscovery of America: Native Peoples and the Unmaking of U.S. History* (New Haven, CT: Yale University Press, 2023). This book came out as mine was ready to go to press. It is invaluable.

16. Michel-Rolph Trouillot, *Silencing the Past: Power and the Production of History* (Boston: Beacon, 1995), 110. Trouillot charts the multimodal silencing of the real story of Columbus through public memorials and official white histories of the Haitian revolution. That study begins to show how whiteness became a political formation. It is suggestive with respect to how other modes of silencing work too, including silencing until late in the day of climate changes rumbling along below and above public attention. The book is remarkable in itself and suggestive for themes pursued in this text.

Chapter 1: Hesiod, Ovid, and a Turbulent Cosmos

1. For a discussion of the invention of this alphabet, see Roderick Beaton, *The Greeks: A Global History* (New York: Basic Books, 2021), 46–52.

2. Hesiod, *Theogony*, trans. and introduced by Norman O. Brown (Saddle River, NJ: Prentice Hall, 1953), 54.

3. *Theogony*, 56–57.

4. In *The Face of the Deep: A Theology of Becoming* (New York: Routledge, 2003), Catherine Keller, a theologian steeped in Jewish and Christian texts, reads the opening lines of Genesis to mean that the tumult of the "deep" preceded God's creation, providing it with both limits and possibilities. This God is not omnipotent, then. This text invites close comparison to the Greek readings and debates we are reviewing.

5. *Theogony*, 58.

6. *Theogony*, 58.

7. *Theogony*, 71.

8. *Theogony*, 72.

9. For this discussion, see Mary Daly, "The Metapatriarchal Journey of Exorcism and Ecstasy," in *The Mary Daly Reader*, ed. Jennifer Rycenga and Linda Barufaldi (New York: New York University Press, 2017), 131–62.

10. *Works and Days*, in *Hesiod: Theogony, Works and Days, Shield*, trans. and ed. Apostolos Athanassakis (Baltimore, MD: Johns Hopkins University Press, 2004), 59–111.

11. I find the fascinating book by Jenny Strauss Clay, *Hesiod's Cosmos* (Cambridge: Cambridge University Press, 2003), to provide an excellent guide here, and in other respects as well.

12. *Works and Days*, 81.

13. *Works and Days*, 77.

14. See Marija Gimbutas, *The Living Goddesses*, ed. Miriam R. Dexter (Berkeley: University of California Press, 2001). For instance, in an early period sculpture, "Hera can be seen standing over him [Zeus] with her arm raised like a great goddess, while he sits in the suppliant son-lover pose" (159–60). I thank Nidesh Lawtoo for calling this text to my attention.

15. We will encounter Carlo Rovelli, a renowned quantum theorist, a few times in this book: here; at another juncture when he elaborates a mode of multi-perspectivism

which shows some affinities to the multi-perspectivism of Amazonians; and at yet another time when he supports an image of time that withdraws singularity from it. For the exploration of Anaximander, see Carlo Rovelli, *Anaximander and the Birth of Science*, trans. Marion Lignana Rosenberg (New York: Riverhead Books, 2021). While Rovelli's own explorations of perspectivism and time seem fecund to me, I am less taken with the short western history of the science/religion struggles he offers in *Anaximander.* It seems to overlook times, some of which will be reviewed in this text, in which certain *complementarities* between the religion and the science of the day blocked needed explorations from proceeding.

16. Tim Whitemarsh, *Battling the Gods: Atheism in the Ancient World* (New York: Alfred A. Knopf, 2015). Pages 92 to 96 are the ones I am summarizing now. But the entire text is important to the issues we are exploring.

17. Whitemarsh, 92.

18. Whitemarsh, 95.

19. Charles Martindale, in his introduction to *Ovid Renewed: Ovidian Influences on Literature and Art from the Middle Ages to the Twentieth Century*, ed. Charles Martindale (Cambridge: Cambridge University Press, 1988), 15.

20. Ovid, *Metamorphoses*, trans. Charles Martin (New York: W. W. Norton, 2004), 15.

21. *Metamorphoses*, 15.

22. Bill McGuire, *Waking the Giant: How a Changing Climate Triggers Earthquakes, Tsunamis and Volcanoes* (Oxford: Oxford University Press, 2012), 3–4.

23. *Metamorphoses*, 15.

24. *Metamorphoses*, 16.

25. *Metamorphoses*, 26–27.

26. *Metamorphoses*, 28.

27. David Montgomery, *The Rocks Don't Lie: A Geologist Investigates Noah's Flood* (New York: W. W. Norton, 2012).

28. *Metamorphoses*, 29.

29. *Metamorphoses*, 103.

30. *Metamorphoses*, 104.

31. *Metamorphoses*, 245.

32. *Metamorphoses*, 247.

33. *Metamorphoses*, 249.

34. Kyle Harper, *The Fate of Rome: Climate, Disease, and the End of an Empire* (Princeton, NJ: Princeton University Press, 2017).

35. *Metamorphoses*, 104.

36. *Metamorphoses*, 105.

37. *Metamorphoses*, 106.

38. See William E. Connolly, "Race and the Anthropocene: Planetary Circuits of Imperial Power," *The Contemporary Condition* (blog), December 14, 2021, http://contemporarycondition.blogspot.com/2021/12/race-and-anthropocene-planetary.html.

39. *Metamorphoses*, 191.

40. *Metamorphoses*, 194–95.

41. See Michel Serres, *The Birth of Physics*, trans. Jack Hawkes (New York: Clinamen, 2001); Laura Tripaldi, *Parallel Minds: Discovering the Intelligence of Materials* (Cambridge, MA: MIT Press, 2021); Jane Bennett, *Vibrant Matter: A Political Ecology of Things* (Durham, NC: Duke University Press, 2010). Serres explores how the Greek sciences of liquids and viscosities were at odds with later sciences that became so enamored with solids as the template of things. Laura Tripaldi explores sciences of complexity, resilience, and emergence that grow out of an exploratory chemistry that has sat in the wings too long. Bennett explores supple images and practices that undermine a metaphysics of solids. Tripaldi explicitly enters into conversation with Ovid. The other two easily could. I entered these conversations most actively beginning with *A World of Becoming* (Durham, NC: Duke University Press, 2011).

42. For one study that explores the capacities of plants for consciousness, planning, feeling, and decision see Paco Calvo with Natalie Lawrence, *Planta Sapiens: The New Science of Plant Intelligence* (New York: W. W. Norton, 2022). The authors suggest that the vascular systems of plants, carrying electrical circuits, loosely serve functions with affinities to those of the neuronal systems of animals. They concur that more research is needed to establish this theory. But think about it the next time you encounter a Venus flytrap.

43. Val Plumwood, *The Eye of the Crocodile*, ed. Lorraine Shannon (Canberra: Australian National University E Press, 2012). Plumwood faced the chilling experience of a crocodile pulling her down into a death roll and then, miraculously, letting go of her halfway down. Surviving, she wrote about how each of us is a potential source of edibility to others. She developed a philosophy of respect for diverse kinds of creatures out of that experience. Her book begins with a quotation from an essay I had written about how you and I are indifferent to the eye of a crocodile, unless and until it looks upon us as potential food. So I have felt connected to her themes for years, though I have so far avoided the crocodile death roll. I also construe her and Ovid to be kissing cousins.

44. *Metamorphoses*, 530.

45. See *The Pythagorean Sourcebook and Library*, comp. and trans. Kenneth Guthrie (Grand Rapids, MI: Phanes, 1987). Here is what Ocellus Lucanus, a Pythagorean writing after Aristotle, says: "Therefore it appears to me that the universe is indestructible and unbegotten, since it always was and always will be; for if it had a temporal beginning, it would not always have existed" (203). And as he makes clear, it could be "corrupted." The beginning of the Ovid cosmos does not sit easily with chapter 15 of his book, unless "permanence" there merely means long lasting. The Roman Empire did last only a few hundred years after Ovid wrote; of course, American capitalism is a baby by that standard.

46. Martindale, *Ovid Renewed*. This is a rich collection of essays exploring medieval and modern reactions to Ovid.

47. Daniel Heller-Roazen, *The Fifth Hammer: Pythagoras and the Disharmony of the World* (New York: Zone Books, 2011).

First Coda: Jocasta, James Baldwin, and Tragic Possibility

1. Thucydides, *The History of the Peloponnesian War*, trans. Richard Crawley (New York: Compass, 2019), 51–53.

2. I find Bernard Williams's comparison of Sophocles and Thucydides in *Shame and Necessity* (Berkeley: University of California Press, 1993) to be perspicuous. One emphasizes preordainment by capricious gods; the other chancy conjunctions. But by comparison to Aristotle, Kant, and Hegel, neither thinks the world is reliably attuned to human aspirations. Mill and Rawls could be added to the first list. "If we reject the progressivist image, however, we shall be more open to the thought that the important question . . . is whether or not a given writer or philosophy believes that beyond some things human beings have themselves shaped, there is anything at all that is intrinsically shaped to human interests. In the light of that question Plato, Aristotle, Kant, Hegel are all on the same side. . . . Sophocles and Thucydides, by contrast, are aligned in leaving us with no such sense. . . . In this perspective the difference between a Sophoclean obscurity of fate and Thucydides' sense of rationality at risk to chance is not so significant" (164–65).

3. Sophocles, *Oedipus the King*, in *The Oedipus Plays of Sophocles*, trans. Paul Roche (New York: Penguin, 1996), 6. I choose this translation partly because I know my way around it, having taught it for several years, and partly because the translator is alert to the cadences and rhythms of the text. Sophocles thinks that cadence and rhythm play major roles in insinuating themes into the psyche. I concur.

4. *Oedipus*, 14–15.

5. *Oedipus*, 40.

6. *Oedipus*, 52. Here is the translation of this key passage from David Grene, the translator held by some to be the most reliable:

Why should man fear when chance is all in all

> For him, and he can clearly foreknow nothing?
> Best to live lightly, as one can unthinkingly.
> As to your mother's marriage bed, don't fear it.
> Before his, in dreams too, as well as oracles,
> Many a man has lain with his own mother.
> But he to whom such things are nothing bears

His life most easily."

Sophocles, *Oedipus the King*, in *Sophocles*, vol. 2 of *The Complete Greek Tragedies*, trans. David Grene, ed. David Grene and Richmond Lattimore (Chicago: University of Chicago Press, 1992), 52.

7. *Time in Greek Tragedy* (Ithaca, NY: Cornell University Press, 1968), by Jacqueline de Romilly, is a powerful book in this respect. She reviews how the image of time shifts as tragic poetry proceeds from the time of Aeschylus, through Sophocles, to Euripides. She is particularly alert to how time speeds up in Sophocles from time to time, as unexpected conjunctions between disparate trajectories erupt. In Sophocles, "time is no longer a coherent sequence in which we painfully achieve the designs of

transcendent justice [as in Aeschylus]: it becomes a series of rough, startling changes, which may occur in any manner, affecting both people's fortunes and feelings" (89).

8. *Oedipus*, 58.

9. *Oedipus*, 60.

10. *Oedipus*, 64.

11. *Oedipus*, 67.

12. *Oedipus*, 80.

13. *Oedipus*, 80.

14. Georg Simmel, in *The Sociology of Georg Simmel*, ed. Kurt Wolff (New York: Free Press, 1962), 145.

15. James Baldwin, *Another Country* (New York: Vintage, 1960), 28.

16. In *James Baldwin's* Another Country (New York: Ig, 2021), Kim McLarin charts and critiques elements of misogyny in that novel. Are they, she asks, set in the voice of Baldwin or in characters he identifies? She suggests he is writing before feminism exercised such a wide influence, but she also knows that in situations of stratification and degradation revenge is apt to take a variety of forms.

17. Baldwin, *Another Country*, 28.

18. *Another Country*, 86.

19. James Baldwin, *Notes of a Native Son* (Boston: Beacon), 89–90.

20. James Baldwin, *The Fire Next Time* (New York: Vintage Books, 1962), 39–40.

21. *Fire Next Time*, 42–43.

22. *Fire Next Time*, 86.

23. *Fire Next Time*, 91.

24. *Fire Next Time*, 91.

25. *Fire Next Time*, 92.

26. *Fire Next Time*, 42–43.

Chapter 2: Augustine and the First Conquest of Pagans

1. Augustine, *The City of God Against the Pagans*, trans. R. W. Dyson (Cambridge: Cambridge University Press, 1998), 55.

2. *City of God*, 56.

3. *City of God*, 233.

4. *City of God*, 145.

5. *City of God*, 145.

6. *City of God*, 242.

7. *City of God*, 244.

8. *City of God*, 249.

9. *City of God*, 250.

10. *City of God*, 881.

11. *City of God*, 267.

12. In the next few paragraphs I draw sustenance from chapter 2 of William E. Connolly, *The Augustinian Imperative: A Reflection on the Politics of Morality* (Newbury Park, CA: Sage, 1993).

13. Augustine, *Select Letters*, trans. James Houston Baxter (Cambridge, MA: Harvard University Press, 1930), Epistle 211, 381.
14. *Select Letters*, Epistle 211, 389.
15. *City of God*, 891.
16. *Selected Letters*, Epistle 232, 469–71.
17. *City of God*, 42.
18. *City of God*, 891–92.
19. *Select Letters*, Epistle 191, 339.
20. Augustine, *The Confessions of St. Augustine*, trans. John Kenneth Ryan (New York: Image Books, 1960), 89.
21. *Confessions*, 166–67.
22. *Confessions*, 197.
23. *Confessions*, 167.
24. *Confessions*, 197.
25. For a fascinating reading of *Antigone*, in which Ismene first resists her sister's call to bury their defeated brother and then secretly does so, see Bonnie Honig, *Antigone, Interrupted* (Cambridge: Cambridge University Press, 2013).
26. *The Book of J*, trans. David Rosenberg and interpreted by Harold Bloom (New York: Grove Weidenfeld, 1990).
27. *Confessions*, 357.
28. Pope Francis, *Laudato Si': On Care for Our Common Home* (Huntington, IN: Our Sunday Visitor, 2015), 57, 62, 79, 107.
29. *Confessions*, 287–88.
30. *Confessions*, 293.
31. Catherine Nixey, *The Darkening Age: The Christian Destruction of the Classical World* (New York: Macmillan, 2017), 93.

Second Coda: Catherine Keller and Diverse Christianities

1. Carlo Ginzburg, *The Cheese and the Worms: The Cosmos of a Sixteenth-Century Miller*, trans. John Tedeschi and Anne C. Tedeschi (London: Routledge and Kegan Paul, 1980), 50–51. This is a superb book, which eventually compares the experiences of Menocchio to several other dissidents of the day.
2. In *The Religion of Jesus the Jew* (Minneapolis, MN: Fortress, 1993), Geza Vermes enumerates several of these differences, including Paul's promise that the second coming was imminent. Here is one thing Vermes says: "A great challenge, perhaps the greatest of them all, which traditional Christianity of the Pauline-Johannine variety has therefore still to confront, does not come from atheism, or agnosticism, or sheer materialism but from within, from the three ancient witnesses, Mark, Matthew and Luke, through whom speaks the chief challenger, Jesus the Jew" (215). This statement, on the last page of the text, summarizes the challenge Vermes advances.
3. Paul Ricœur, *The Symbolism of Evil*, trans. Emerson Buchanan (Boston: Beacon, 1967).

4. In *Black Prophetic Fire*, ed. Christa Buschendorf (Boston: Beacon, 2014), Cornel West measures his debts and differences with a host of Black activists, including Martin Luther King, Frederick Douglass, W. E. B. Du Bois, and Ella Baker. His own conjunctions between activism, deep pluralism, and egalitarianism are indispensable. They invite comparisons with Catherine Keller, too, who forms the subject of this Coda. Indeed, the two entered into conversations several years ago at an APSA meeting when both assessed my work in *Capitalism and Christianity, American Style* (Durham, NC: Duke University Press, 2008).

5. Charles Taylor, *A Secular Age* (Cambridge, MA: Harvard University Press, 2007).

6. See *The Mary Daly Reader*, ed. Jennifer Rycenga and Linda Barufaldi (New York: New York University Press, 2017).

7. Catherine Keller, *Political Theology of the Earth: Our Planetary Emergency and the Struggle for a New Public* (New York: Columbia University Press, 2018), 12.

8. See Catherine Keller, *On the Mystery: Discerning Divinity in Process* (Minneapolis, MN: Fortress, 2008). See in particular 269–73.

9. *The Book of J*, trans. David Rosenberg and interpreted by Harold Bloom (New York: Grove Weidenfeld, 1990), 61.

10. Quoted in Catherine Keller, *The Face of the Deep: A Theology of Becoming* (New York: Routledge, 2003), 114.

11. *The Book of J*. So the contention of Bloom and Rosenberg is that Augustine, among others, misread the early Hebrew upon which the Christian version of Genesis was anchored.

12. Keller, *Face of the Deep*, xv.

13. See William James, *A Pluralistic Universe* (Lincoln: University of Nebraska Press, 1996).

14. Keller, *Face of the Deep*, 198.

15. Keller, *Political Theology of the Earth*, 167.

16. Several texts correct earlier stories in which enslavement was limited only to Africans and territorial dispossession to Amerindians. Amerindians were often enslaved, as, for instance, Ned Blackfoot records in *The Rediscovery of America: Native Peoples and the Unmaking of U.S. History* (New Haven, CT: Yale University Press, 2023). And territorial dispossession, it must not be forgotten, became a scourge in Africa. Moreover, the white constituencies who defended African enslavement intransigently were often those who pressed for Amerindian dispossession most intensely.

17. Bill McGuire, *Waking the Giant: How a Changing Climate Triggers Earthquakes, Tsunamis and Volcanoes* (Oxford: Oxford University Press, 2011), 38–39.

18. Keller, *Political Theology of the Earth*, 160, 161.

19. Keller, 168.

20. Keller pursues another opening, too, worthy of note here. In *Facing Apocalypse: Climate, Democracy, and Other Last Chances* (Maryknoll, NY: Orbis Books, 2021) she mines the Book of Revelation to explore the sense of planetary

volatility there connected to a punishing, vengeful god. It may well be that her extractions can be brought into productive comparison with those I am about to draw from a few pagan cultures. I have not yet pursued that trail.

Chapter 3: Todorov, the Second Conquest, and Aztec Cosmology

1. For the higher estimate see Henry Dobyns, "Estimating Aboriginal Populations: An Appraisal and Techniques with a New Hemispheric Estimate," *Current Anthropology* 77, no. 3 (October 1966): 395–449. And for the lower see Suzanne Austin Alchon, *A Pest in the Land: New World Epidemics in a Global Perspective* (Albuquerque: University of New Mexico Press, 2003).

2. Tzvetan Todorov, *The Conquest of America: The Question of the Other*, trans. Richard Howard (New York: Harper and Row, 1984), 20.

3. Todorov, 21.

4. Todorov, 27.

5. Todorov, 50. In *Silencing the Past* (Boston: Beacon, 1995), Michel-Rolph Trouillot reviews how Columbus, primarily oriented to the Caribbean islands and seeking a sea route to the East, became by the 1890s a full-fledged hero to burgeoning settler populations of Italians and Irish in the United States. This also helped lift those two demeaned immigrant populations into the status of full-fledged whites, that is, to complete the political formation of whiteness in North America. Columbus Day eventually became a patriotic holiday designed to monumentalize white, Christian America, as well as to invite Catholics into the sanctified identity.

6. See Jonathan Kennedy, *Pathogenesis: A History of the World in Eight Plagues* (New York: Crown, 2023). This book offers a corrective to sociocentric variants of the human sciences. It strives to trace the role of pathogenesis in human societies from the earliest to most recent times. The discussion of Columbus and the introduction of African enslavement into the Caribbean is on pages 154–60. It is also true perhaps that the book short circuits some of the positive horizontal exchanges between viruses and bacteria on the one side and the human estate on the other that have helped to forge human cultures.

7. For a book that begins to explore some of these dynamics, see Philipp Blom, *Nature's Mutiny: How the Little Ice Age of the Long Seventeenth Century Transformed the West and Shaped the Present* (New York: Liverwright, 2019).

8. Todorov, *Conquest*, 56.

9. For one fascinating and recent study see Camilla Townsend, *Fifth Sun: A New History of the Aztecs* (Oxford: Oxford University Press, 2019). Townsend finds the story of Montezuma's prophecy to be apocryphal, basing her evidence on texts written by indigenous authors and Cortés himself. Her account of the invasion, early defeats by the invaders, the effects of rolling diseases on the locals, and the awareness of Montezuma that new waves of invaders would soon follow surely improves upon the Todorov account. For another recent account see Matthew Restall, *Seven Myths of the Spanish Conquest*, rev. ed. (Oxford: Oxford University

Press, 2021). He concludes that disease (with differential immunities to it), a nonunified region upon contact, advantages provided by horses and dogs in combat, steel swords, and a Spanish culture of war and superiority played major roles; several of these factors evened out after the initial conquest. The persisting value of the Todorov text, in the face of these corrections, resides in its accounts of diverse struggles by invading priests to square the incredible violences of conquest with Christian doctrine and, sometimes, to modify the latter a bit, though not enough to demand a pullout. Townsend and Restall, unlike Todorov, also explore struggles of resistance and patterns of creolization that began to emerge a generation after the conquest. The conquest, that is, was partial. In this way Townsend's book is comparable to that of Ned Blackhawk, who explores twentieth-century patterns of resistance in North America: *The Rediscovery of America: Native Peoples and the Unmaking of U.S. History* (New Haven, CT: Yale University Press, 2023).

10. I have explored those issues in two books: Connolly, *Identity/Difference: Democratic Negotiations of Political Paradox* (Ithaca, NY: Cornell University Press, 1991); and *The Ethos of Pluralization* (Minneapolis: University of Minnesota Press, 1995). The first book includes my first encounter with Todorov, one that made a lot out of how western imperialism helped define the field of "international relations." But it did not focus on how the pagan gods the Spanish encountered had things to teach them and us.

11. Todorov, *Conquest*, 164.

12. Todorov, 245.

13. Blackhawk, *Rediscovery of America*, 356.

14. For a superb account of how creolization works in the Caribbean today, as well as an exemplification of how it functions as an ideal in his own work, see Édouard Glissant, *Poetics of Relation*, trans. Betsy Wing (Ann Arbor: University of Michigan Press, 2010).

15. Todorov, *Conquest*, 227.

16. Todorov, 241.

17. A reflective account of the role of plagues during several historical turns can be found in Jonathan Kennedy, *Pathogenesis: A History of the World in Eight Plagues*. He would probably embrace the comparison between Rome and the Aztecs made above, though I would paint its tones somewhat differently. These mass conversions, while never total and replete with qualifications, may have led to cultural depreciation of respect for the shaky place of the human estate in the cosmos in some pagan orientations operative in Greece, Rome, and Mexico.

18. Tzvetan Todorov, *On Human Diversity: Nationalism, Racism, and Exoticism in French Thought*, trans. Catherine Porter (Cambridge, MA: Harvard University Press, 1993), 390.

19. Todorov, 390.

20. Todorov, 394. The latest book I have encountered by Todorov was published in 2006. It does not move on any of the fronts identified here to be problematic. It may be wise to refrain from criticizing him in these respects, merely noting

assumption and priorities he shared with so many others. That is probably right. On the other hand, he did have potential experience with Aztec cosmology to consult; perhaps it could have enabled him to prod some of the assumptions he shared with so many others in the west. See *In Defence of the Enlightenment*, trans. Gila Walker (London: Atlantic Books, 2006).

21. My first engagement with the trickster was in *Politics and Ambiguity* (Madison: University of Wisconsin Press, 1987). The task accepted was neither to rise above trickster myths nor to embrace them exactly as they were. It was to try to pull some things from those practices that could qualify western cultural images of mastery of nature and/or harmonious attunement to a nature that is smooth in itself. This quotation perhaps captures something of the gist of that effort: "The trickster, a protean self whose multiple centers of action often contend with one another, acknowledges a struggle others tend to suppress" (139). That 1987 text begins to join the sense of the self as multiplicity to that of a planet that is rockier than many western mastery and harmony problematics concede. But it did not yet push the latter theme very far.

22. Bernardino de Sahagún, *A History of Ancient Mexico*, trans. Fanny R. Bandelier (New York: Fisk University Press, 1932), 25.

23. Sahagún, 25.

24. Sahagún, 26.

25. Sahagún, 26.

26. Sahagún, 29.

27. Sahagún, 30.

28. See Paul Radin, *The Trickster: A Study in American Indian Mythology* (New York: Schoeken Books, 1967).

29. Sahagún, *History of Ancient Mexico*, 182.

30. Frances F. Berdan, *The Aztecs: Lost Civilizations* (London: Reaktion Books, 2021). This book updates and corrects several assumptions Todorov made about the Aztecs, drawing upon more recent research.

31. Guilhem Olivier, *Mockeries and Metamorphoses of an Aztec God: Tezcatlipoca, "Lord of the Smoking Mirror,"* trans. Michel Besson (Boulder: University Press of Colorado, 1997), 137. The god Olivier focuses on is Tezcatlipoca, the one able to stop the sun and start new eras in the roughly cyclical image of the cosmos that governs the Aztecs. I say "roughly" because it seemed possible to delay or hasten these cycles through anger of the gods, by helping them, and by sacrificing to them. That, anyway, is what I think Olivier means when he says "it is *primarily* the signs under which they have born that determine their valor or their cowardice and their predisposition to meet a particular destiny" (18, my emphasis).

32. Some support for the above paragraph may be found in Eduardo Viveiros de Castro, *The Inconstancy of the Indian Soul: The Encounter of Catholics and Cannibals in 16th-Century Brazil* (Chicago: Prickly Paradigm, 2011). It must be emphasized from the start that he is exploring sixteenth-century priestly orientations to Amazonians in what is now Brazil, not the Aztecs. Nonetheless, quotations such as this encourage

me to project the above comparisons between Augustinians and Aztecs. "I am not saying . . . that something like a religion may not have existed, or a cultural order, or a Tupinamba society: I am only suggesting that this religion was not framed in terms of the category of belief. . . . What I am saying is that Tupinamba philosophy affirmed an essential ontological incompleteness; the incompleteness of society and, in general, of humanity" (47). What might be discerned here is both a connection to Augustinian belief and a difference from it, two different crystallizations that cross into one another. Hence, one additional source of the need to complicate binary logics in cultural interpretation.

33. In a very helpful reply to an email I had addressed to him, James Maffie clarified his current notion of the Aztec gods and corrected some mistakes I had made on other fronts. I hope the latter have been now corrected. I retain, however, the idea that it can be productive to compare the Aztec gods and cosmology to Greek cosmologies. The idea would be not to find *identities* here but to locate *affinities* across significant differences that may help critics today better understand why western Christian/secular cultures both produced the time of climate wreckage and took soooo long to perceive it. I understand why Maffie is cautious here. It is treacherous territory. But if climate wreckage is one of the key issues today, that caution need not perhaps become a refusal. I do expect that some of my extrapolations will be in need of further correction.

34. James Maffie, *Aztec Philosophy: Understanding a World in Motion* (Boulder: University Press of Colorado, 2014), 468 and 482.

35. Maffie, 378.

36. Maffie, 384.

37. Maffie, 523. Maffie begins his book by arguing, against Husserl, Russell, Rorty, and others, that the Aztecs did think philosophically. Fundamental metaphysical speculation is not confined to the west, though, I add, those who have been sunk in the quagmires of human exceptionalism and planetary gradualism have been tempted to deny this. Since many valuable Aztec texts were burned by Christian/conquistador invaders, Maffie builds his case through engagements with Aztec art and sculpture, texts by Sahagún and Duran, and contemporary Nahuatl practices. He shows how these practices break with Christian metaphysics, Platonic metaphysics, and secular metaphysics of the west. I only wish he had compared his process interpretation to that of a process metaphysics today, selecting either James, Bergson, Nietzsche, Whitehead, Deleuze, or Serres for comparison. My own sense is that Nietzsche comes close at least in thinking about how the order/disorder of the cosmos makes human life fragile (and encourages many practices of existential resentment), and that Serres moves close too in construing time to be composed of multiple temporalities that both collide and do not always coalesce. We will encounter both of the latter in later parts of this text. I feel confident, on the other hand, that Maffie would wish that I was more subtle in my review of the Aztecs.

Third Coda: Tocqueville and White Settler Society

1. See, for instance, Michael Hardt, ed., *Thomas Jefferson: The Declaration of Independence* (New York: Verso, 2007). The letters are collected in a section titled "Native Americans and Black Slavery." Here is one example: "Comparing them by their faculties of memory, reason, and imagination, it appears to me that in memory they are equal to whites; in reason much inferior . . . ; and that in imagination they are dull, tasteless, and anomalous" (80).

2. Alexis de Tocqueville, *Democracy in America*, trans. George Lawrence (New York: Anchor Books, 1969), 355.

3. My first engagement with Tocqueville on Amerindians was in Connolly, "Tocqueville, Territory, and Violence," *Theory, Culture & Society* 11, no. 1 (February 1994): 19–42. That essay then sought to imagine a new pluralism that proceeds well beyond the confined civi-territorial mode propagated by Tocqueville. Here, while retaining that objective, I also seek to underline the ways Tocqueville's intertwining of territory, race, property, and Christianity both delineated the limits within which his thinking was set and helped to sow the seeds of future climate wreckage. Those plowed fields (and deforestation) he loved so much formed part of that story.

4. Tocqueville, *Democracy in America*, 27–29.

5. Tocqueville, 292. Michel Rolf Trouillot, in *Silencing the Past: Power and the Production of History* (Boston: Beacon, 1995), explores why creolization and blending succeeded better in Latin America than in North America, even though it seldom reached into the centers of state power there. The text is also superb in showing how so much of the history we are touching has been silenced, not only by power elites but also by the very cosmological assumptions taken for granted in the governing regimes.

6. For a fine book that adumbrates how these orientations became encoded into white settler memory, see Kevin Bruyneel, *Settler Memory: The Disavowal of Indigeneity and the Politics of Race in the United States* (Chapel Hill: University of North Carolina Press, 2021). Bruyneel shows how settler memory of Bacon's rebellion in 1675 erases the initial alliance between white laborers, the enslaved, and Amerindians. He also contends that some critical memories of the same events involve disavowal of the Amerindian holocaust while focusing on the resulting separation between Whites and Blacks. Joel Olson, for instance, in *The Abolition of White Democracy* (Minneapolis: University of Minnesota Press, 2004), begins by noting how the Bacon event helped to construct whiteness by erasing identifications between lower-class whites, enslaved Africans, and Amerindians; he then drops Amerindians from the analysis in focusing on White/Black enmities. This is a mode of disavowal. The Bruyneel book is an impressive account of the formation and disavowals of white settler memory in the United States. I do wish he had tended more to the role of Protestant Christianity in that fateful formation. A final entry into this discussion is important, too. P. J. Brendese, in *Segregated Time* (New York:

Oxford University Press, 2023), charts closely how dominant white experiences of time devalue the lived experiences of many African Americans and Amerindians, with serious consequences for the unconscious partitions of life. This book suggests the need to adumbrate another image of time, a task I will take up in the last chapter.

7. Barry O'Connell, ed., *On Our Own Ground: The Complete Writings of William Apess, a Pequot* (Amherst: University of Massachusetts Press, 1992), 286–87. But, of course, few whites encountered the veritable Apess reply until O'Connell, a professor at Amherst College, collected and published these essays in 1992. O'Connell, as he made transparent to readers and colleagues, taught in a town and at a college named in honor of Lord Jeffery Amherst, a renowned Indian fighter. I taught in that town for several years too, at a different school.

8. For a brilliant account that updates Apess and much more, see Ned Blackhawk, *The Rediscovery of America: Native Peoples and the Unmaking of U.S. History* (New Haven, CT: Yale University Press, 2023). In one chapter Blackhawk shows how the weakness of divided Confederate states with respect to historic Amerindian peoples helped to foster new drives for a more unified federalist state. His accounts of how "whiteness" became formed and destructive rampages of white settler militia groups speak to both that time and today. The militia tradition is long and remains entrenched in white rural strongholds.

9. See Sam White, *A Cold Welcome: The Little Ice Age and Europe's Encounter with North America* (Cambridge, MA: Harvard University Press, 2017), 105. This book also reviews the effects of climate change on Amerindians and invaders in the western part of the current United States.

10. Roxanne Dunbar-Ortiz, *An Indigenous Peoples' History of the United States* (Boston: Beacon, 2014), 63.

11. Numerous formulations such as this recur, exemplified by an account of a white woman who escaped an Indian "uprising" in Minnesota. "Help finally came [to the white lad] in the form of his erstwhile dying mother who had at last found new strength to rise up and start out in the direction of civilization" (Ralph Andrist, *The Long Death: The Last Days of the Plains Indians* [New York: Collier Books, 1964], 51). It is one thing to appreciate the heroism of the mother, another to enclose it unconsciously in "civilization," as if the Sioux peoples *lacked* such a *singular* achievement. The language makes the "uprising" and the response regrettable while softening the sense that the combination is the result of white colonial invasions of the lands of others. A more recent text better emphasizes how long resistances occurred and how a profound legacy of Indigenous civilization persists. See Pekka Hämäläinen, *Indigenous Continent: The Epic Contest for North America* (New York: W. W. Norton, 2022). There are many riches in this text, including, for instance, the impressive account of how a series of peoples—the Lakota, the Comanche, the Apache—revolutionized their own modes of life during a short period when they constructed new ways of life around the horse. Only the construction of the railroad and Morse code by the white settler state were eventually able to overcome the speed, resilience, and adaptability that this new mode of being enabled.

12. For an effective history of the frontier myth, its power, and reactions to its demise, see Greg Grandin, *The End of the Myth: From the Frontier to the Border Wall in the Mind of America* (New York: Henry Holt, 2019).

13. Roxanne Dunbar-Ortiz, *Not "A Nation of Immigrants": Settler Colonialism, White Supremacy, and a History of Erasure and Exclusion* (Boston: Beacon Books, 2019), 20.

14. A text in which I explore this phenomenon is Connolly, *Aspirational Fascism: The Struggle for Multifaceted Democracy under Trumpism* (Minneapolis: University of Minnesota Press, 2017). It appeared during the first several months of the Trump regime when some on the Left still assumed it to be a "populist" movement operating roughly within the guardrails of democracy. The January 6 Insurrection exploded that assumption.

15. In an August 14, 2022, commentary titled "The Politics of Persecution," Charles Blow examines how such an inflationary rhetoric works. The specific occasion was the event in which Trump insisted he was being persecuted by the FBI when it arrived at his residence with a court order to recapture highly classified documents he had stolen upon departure from the White House for Mar-a-Lago (*New York Times*, August 14, 2022, SR3). I was pleased that he found an earlier essay by me on this phenomenon to be helpful. Its logic, again, is first to presume that you have a rightful claim to full hegemony as white, or Christian, or straight, or male, etc. And then to cry persecution when others refuse to honor your first assertion of superiority.

Chapter 4: Descartes, Kant, and Amazonian Perspectivism

1. I note two texts here. First, Charles W. Mills, *Black Rights/White Wrongs: The Critique of Racial Liberalism* (New York: Oxford University Press, 2017). He traces, among other things, how the Lockean contract both conceals Black and indigenous exploitations and continues forward into the 1970s and '80s debates between John Rawls and Robert Nozick. The second, Elisabeth R. Anker, *Ugly Freedoms* (Durham, NC: Duke University Press, 2022), explores connections between Lockean contractualism and his commitments to colonialism. Her discussion of how the harsh plantation slavery in Barbados sugar production was consciously transplanted to South Carolina is superb.

2. See Hans Blumenberg, *The Legitimacy of the Modern Age*, trans. Robert Wallace (Cambridge, MA: MIT Press, 1983), 171, 172.

3. Willard von Quine and J. S. Ullian, *The Web of Belief*, 2nd ed. (New York: Random House, 1978).

4. René Descartes, *Discourse on Method and Meditations*, trans. Laurence J. Lafleur (Saddle River, NJ: Prentice Hall, 1952), 97.

5. Descartes, 97.

6. Quotation from Philipp Blom, *Nature's Mutiny: How the Little Ice Age of the Long Seventeenth Century Transformed the West and Shaped the Present*, translated by author (New York: Liverwright, 2019), 165.

7. We will discuss relations between tactics of the self (gymnastics) and reflexivity later in this text. For now three texts might be consulted: Friedrich Nietzsche, *Writings from the Late Notebooks*, ed. Rüdiger Bittner (Cambridge: Cambridge University Press, 2003); Michel Foucault, "On the Genealogy of Ethics: Overview of Work in Progress," in *The Foucault Reader*, ed. Paul Rabinow (New York: Pantheon Books, 1984), 340–72; Jane Bennett, *Influx and Efflux: Writing Up with Walt Whitman* (Durham, NC: Duke University Press, 2020).

8. See, for example, Antonio R. Damasio, *Descartes' Error: Emotion, Reason, and the Human Brain* (New York: Avon Books, 1994).

9. Descartes, *Meditations*, 105.

10. Descartes, *Discourse on Method*, 45.

11. For discussions of Kant's hypotheses about the Lisbon quake, see Svend Larsen, "The Lisbon Earthquake and the Scientific Turn in Kant's Philosophy," *European Review* 14, no. 3 (July 2006): 359–67; O. Reinhardt and D. R. Oldroyd, "Kant's Theory of Earthquakes and Volcanic Action," *Annals of Science* 40, no. 3 (July 1983): 247–72.

12. For a fine study that compares the Kantian image of time to those offered, among several others, by Hesiod, Aristotle, Plato, Augustine, Nietzsche, and Deleuze, see Philip Turetzky, *Time* (New York: Routledge, 1968). Deleuze, in *Difference and Repetition*, trans. Paul Patton (Minneapolis: University of Minnesota Press, 1994), argues that the subjectivity of temporal experience in Kant implies that time divides between past and present, creating a fracture in the I itself. This, to Deleuze, opens Kantianism potentially to images of time advanced by Bergson and Nietzsche.

13. Immanuel Kant, *Critique of Practical Reason*, trans. Lewis White Beck (New York: Library of Liberal Arts, 1993), 48–49.

14. The section to follow expands in a different key some themes about Kant presented in William E. Connolly, *The Fragility of Things: Self-Organizing Processes, Neoliberal Fantasies, and Democratic Activism* (Durham, NC: Duke University Press, 2013), 100–108. The point then was to focus on Kant's image of morality, its postulate about nature, and how its human exceptionalism almost became qualified in the *Third Critique*. It was also to pose the counterimage of an ethic of cultivation to a morality of universal law. The point now is to show how intersections between the Kantian image of morality, its image of the subject, its necessary postulates about time, and its gymnastics can be placed into productive contestation with an Amazonian multiperspectivism that emphasizes a world composed of multiple, intersecting subjectivities of heterogeneous sorts. To soften the subject/object dichotomy with a world of diverse subjectivities. Doing so to encourage a multiperspectivism. It is also to set the stage for exploration of an alternative image of time to the one Kant postulates. I hope the new account absorbs and broadens the earlier one.

15. Immanuel Kant, "On the Proverb: That May Be True in Theory but Is of No Practical Use," in *Perpetual Peace and Other Essays on Politics, History, and Morals*, trans. Ted Humphrey (Indianapolis: Hackett, 1983), 86.

16. It is a limit of this chapter, as noted in the text, that it does not address the "intuitive" images of space and time articulated in *The Critique of Pure Reason*, though the Kantian postulates discussed here do rest upon them. For an eco-centered book that does address those questions, See Matthew David Segall, *Crossing the Threshold: Etheric Imagination in the Post-Kantian Process Philosophy of Schelling and Whitehead* (Olympia: Integral Imprint, 2023). He draws upon Schelling and Whitehead to challenge Kant. And the quest is rich. It does, however, pursue its counterpoint to a mechanistic world by invoking a series of world harmonies. My attempt, of course, is to challenge *both* of these orientations, as became clear in the readings of Aztec and Hesiod cosmologies.

17. Kant, "To Perpetual Peace: A Philosophical Sketch," in *Perpetual Peace and Other Essays*, 120.

18. Kant, 120.

19. For an early essay supporting the empirical probability of a racial and civilizational hierarchy, see Kant, "Of the Different Human Races," in *The Idea of Race*, ed. Robert Bernasconi and Tommy Lee Lott (Indianapolis: Hackett, 2000). Some debate whether the later Kant held such views (at that point as postulates, not as mere observations), but the case in favor of concluding he did is rather strong. See for example *Anthropology from a Pragmatic Point of View*, trans. Victor Dowdell (Carbondale: Southern University Press, 1978), a text finished late in the day, where he says "Nature has preferred to diversify infinitely the characters of the same stock" (237). Given the necessary postulate of universal, progressive time, this diversity will be seen through the lens of civilizational superiority/inferiority. In a way, the early Kant account is preferable to the later one, since it could be corrected through experience. Moral universalism tied to a cleansed Christianity poses a more serious issue. It lodges civilizational and racial judgments more closely in the march of time itself.

20. See William E. Connolly, "'Beyond Good and Evil': The Ethical Sensibility of Michel Foucault," *Political Theory* 21, no. 3 (August 1993): 365–89.

21. Kant, *Critique of Practical Reason*, 153.

22. Eduardo Viveiros de Castro, *The Relative Native: Essays on Indigenous Conceptual Worlds* (Chicago: Hau Books, 2015), 197. In this brief presentation of Viveiros de Castro's thought I do not dig into his later, explosive book, *Cannibal Metaphysics*, ed. and trans. Peter Skafish (Minneapolis: University of Minnesota Press, 2017), though it would certainly be pertinent to do so. There the role of shamans in enabling communication across human/animal perspectives is highlighted and Viveiros de Castro's relations to Deleuze and Lévi-Strauss are explored. That is for another time.

23. For a reflective set of comparisons between Viveiros de Castro, Philippe Descola, and Bruno Latour on subjectivities, multinaturalism, and perspectivism during the time of the Anthropocene, see Eduardo Kohn, "Anthropology of Ontologies," in *Annual Review of Anthropology* 44 (August 2015): 311–27. Kohn responds effectively to the charge that such approaches are apolitical, locating their

emergence in the west wisely during the shock of the Anthropocene. And he notes how different perspectives can disturb and challenge one another, thereby distinguishing that doctrine from relativism. Multiperspectivism, however, is generally set, as he does, in a multiplicity of beings and subjectivities, human and nonhuman. The essay does not cope with the difficult issue of whether there are also nonhuman *forces* that are not exactly subjects but do not correspond either to traditional western images of objects. I explore tentatively that issue below, through an engagement with one version of quantum mechanics.

24. For one attempt to construct such a broadened coherence theory of truth, drawing the work of Whitehead and Foucault closer together, see, William E. Connolly, *Climate Machines, Fascist Drives, and Truth* (Durham, NC: Duke University Press, 2019), chap. 3, "The Lure of Truth."

25. Friedrich Nietzsche, *The Will to Power*, trans. Walter Kaufmann and R. J. Hollingdale (New York: Vintage Books, 1967), 339–40. My italics.

26. Carlo Rovelli, *Helgoland: Making Sense of the Quantum Revolution*, trans. Erica Segre and Simon Carnell (New York: Riverhead Books, 2020), 182, 85, 88, 154, 88.

27. For a fascinating text that explores empirically and speculatively the emergence of life from nonlife, see Terrence W. Deacon, *Incomplete Nature: How Mind Emerged from Matter* (New York: W. W. Norton, 2012). Deacon contends that the formation of two kinds of complex molecules formed the basis for life, but its emergence required a chancy conjunction between them. So both complexity and chance play a role in his theory. He shares with Rovelli the understanding that quantum theory has demolished the assumption of any primary, unchangeable objects—atoms, particles, or photons—which could form the building blocks of a non-probabilistic theory.

28. Rovelli, *Helgoland*, 78.

Fourth Coda: Nietzsche and the History of an Error

1. In *Old Gods, New Enigmas* (New York: Verso Books, 2016), the Marxist Mike Davis brings the work of Prince Kropotkin, the utopian socialist often maligned during Marx's day, to help unravel the planetary gradualism that limited Marx and Marxists for too long. Kropotkin's was an early attempt to "make a comprehensive case for natural climate change as a prime mover in the history of civilizations. As noted earlier, early Enlightenment and Victorian thought had universally assumed that climate was historical stable, stationary in trend, with extreme events as simple outliers of a mean state" (182). Nietzsche and Kropotkin concur in challenging this assumption. Davis's own work draws that assumption into Marxist theory, as becomes so clear in *Late Victorian Holocausts* (2002). For an essay that appreciates Davis on this issue see William E. Connolly and Jairus Grove, "Planetary Events, Climate Catastrophes, and the Limits of the Human Sciences," in *Between Catastrophe and Revolution: Essays in Honor of Mike Davis*, ed. Daniel Monk and Michael Sorkin (New York: Or Books, 2022), 29–54.

2. Friedrich Nietzsche, *Twilight of the Idols and The Anti-Christ*, trans. R. J. Hollingdale (New York: Penguin, 1968), 40.

3. Nietzsche, 40.

4. Nietzsche, 40.

5. Friedrich Nietzsche, *Writings from the Late Notebooks*, ed. Rüdiger Bittner (Cambridge: Cambridge University Press, 2003), 224n121.

6. Carlo Rovelli, *Helgoland: Making Sense of the Quantum Revolution* (New York: Riverhead Books, 2021), 154.

7. Nietzsche, *Late Notebooks*, 42 (emphases in original).

8. For an essay that reviews Nietzsche's critique of binary logic and begins to explore his relation to perspectivism and multivalued logics, see Steven D. Hales, "Nietzsche on Logic" *Philosophy and Phenomenological Research* 56, no. 4 (December 1996): 819–35. Hales connects this theme to perspectivism. I would add that a perspectivism of the Nietzschean and Rovelli sorts speaks to *both* how this or that entity receives another and how it, often enough, is surprised by it, sometimes opening a potential door to perspectival adjustments. A bacterium adjusts its route when the path to sugar lure is blocked.

9. Bertrand Russell, *The ABC of Relativity* (New York: Harper and Brothers, 1925), 144. In this book Russell explains and defends relativity theory, showing how logical positivism as a philosophy which extends its method and assumptions to all domains of science and social life also helped set an agenda for that theory. Later efforts, showing the limited domain in which relativity applies, also helped foment objections to the moment of positivist hegemony.

10. Nietzsche, *Twilight of the Idols*, 41 (emphases original).

11. See Angela Carter, *The Infernal Desire Machines of Doctor Hoffman* (London: Penguin Books, 1972).

12. Friedrich Nietzsche, *Thus Spoke Zarathustra: A Book for None and All*, trans. Walter Kaufmann (New York: Penguin Books, 1968), 13. I do not present a reading of the "overman" here. Except to say, first, that such a being seeks to overcome existential resentment of death as a condition of life and of the shakiness of the human estate on earth and, second, that on my reading, Nietzsche's take on the overman shifts as *Zarathustra* evolves, until it becomes a more active voice in the self rather than a separate, intact type. Such a reading can be found in William E. Connolly, *A World of Becoming* (Durham, NC: Duke University Press, 2011), chapter 4.

13. Nietzsche, *Twilight of the Idols*, 43 (emphasis original).

14. For an inspired and inspiring book that explores in depth Nietzsche's attachment to the earth, see Henk Manschot, *Nietzsche and the Earth: Biography, Ecology, Politics*, trans. Liz Waters (London: Bloomsbury, 2021). For a dissertation arguing persuasively that Nietzsche challenged logocentrism and human exceptionalism together, while most of his modern followers have adopted only one or the other of those two orientations, see Tvrtko Vrdoljak, "The Metaphysics of Politics: Nietzsche and His Interlocutors" (PhD diss., Johns Hopkins University, 2023).

Chapter 5: Amitav Ghosh, Michel Serres, and the Time of Climate Wreckage

1. Central Calvin themes and disciplines can be found in John Calvin, *The Institutes of Christian Religion*, ed. Tony Lane and Hilary Osborne (Newark, NJ: Baker House, 1986). Weber's classic text is *The Protestant Ethic and the Spirit of Capitalism*, trans. Talcott Parsons (New York: Charles Scribner's Sons, 1958). For a book that assesses Weber's projection that capitalism would henceforth become increasingly secular and then explores the formation of a new "evangelical/capitalist resonance machine," see William E. Connolly, *Capitalism and Christianity, American Style* (Durham, NC: Duke University Press, 2008).

2. In *Climate Machines, Fascist Drives, and Truth* (Durham, NC: Duke University Press, 2019) I present an account of capitalism as an axiomatic with several variations, including neoliberal, Keynesian, and fascist modes of variation. The roles that new shocks (inflation, defeat in war, climate and microbe crossings, etc.) and changes in institutional ethos play are underlined.

3. Amitav Ghosh, *The Nutmeg's Curse: Parables for a Planet in Crisis* (Chicago: University of Chicago Press, 2021), 165 and 167, respectively.

4. Ghosh, *Nutmeg's Curse*, 208, 222, 236, 242, and 251.

5. In *River Life and the Upspring of Nature* (Durham, NC: Duke University Press, 2023), Naveeda Khan explores the relations between such goddesses and gods and the river people of Bangladesh, people who struggle to live on "chars" in the river that may be eroded at one moment and reemerge at another.

6. Bill McGuire, *Waking the Giant: How a Changing Climate Triggers Earthquakes, Tsunamis and Volcanoes* (Oxford: Oxford University Press, 2011), 61–62. As the title of this book suggests, after reviewing the history of previous rocky periods, McGuire, a vulcanologist, presents evidence that the rapid climate shifts of today will increase the number and size of future earthquakes and volcanoes. Volcanoes, after the initial cooling before large sulphur particles fall to earth, seed the sky with yet more greenhouse gases. These are examples of what I call planetary amplifiers.

7. See Michael E. Mann, *Our Fragile Moment: How Lessons from Earth's Past Can Help Us Survive the Climate Crisis* (New York: PublicAffairs, 2023), 222–23. I find Mann's new account persuasive on this count. My worries about the contrast model of doomsdayers governing his narrative are presented later in this chapter.

8. They are presented in McGuire, *Hothouse Earth: An Inhabitant's Guide* (London: Icon Books, 2022).

9. My first effort to chart this process was presented in a post in *The Contemporary Condition* in December 2021, titled "Race and the Anthropocene: Planetary Circuits of Imperial Power," http://contemporarycondition.blogspot.com/2021/12/race-and-anthropocene-planetary.html.

10. Bill McGuire and Michael Mann note the instability of the Thwaites Glacier; my account is also rooted in the more recent essay by Douglas Fox, "The Coming Collapse," *Scientific American* 327, no. 5 (November 2022): 32–41.

11. Some will say this should not be called "imperial power" because the emitters do not intend the effects. I disagree, drawing attention to the massive suffering the emissions carry to other regions and the comparative advantages these effects carry for elites in emitting regimes. But if you think that argument does not suffice, do not forget that these circuits become increasingly known by emitting powers. They thus now become intentional. Impersonal, planetary circuits of imperial power.

12. Here is what Michael Mann, who formerly doubted claims that the ocean conveyor is in a precarious state, now says: "Climate models predict the AMOC will weaken later this century, if we continue burning carbon. . . . The paleo observations spanning the past two millennia beg to differ, however—they suggest that a dramatic slowdown has already begun" (*Our Fragile Moment*, 199). The statement underlines how computer predictive models often underplayed the dynamism and interactions of diverse forces that mark climate volatility.

13. One book that traces numerous such temporal asymmetries is Adam Welz, *The End of Eden: Wild Nature in the Age of Climate Breakdown* (New York: Bloomsbury, 2023).

14. Michael Mann, *Our Fragile Moment.* Here is one of many formulations that appear shortly after the doomsday contrast model is mobilized and rejected: "Dangerous climate change cannot be avoided. It's already here. So it's a matter of how bad we're willing to let it get. . . . The IPCC estimates as much as fourteen percent of species could be lost at 1.5 ° C (2.7 ° F) warming. . . . However, the number climbs to twenty-nine percent at 3 ° C (5.4 ° F), . . . and forty-eight percent at 5 ° C (9 ° F). . . . But that is avoidable in a scenario of meaningful climate action" (232). Failing to coordinate those geological findings with corollary work in social theory on climate events and the twin dangers of casualism and fascism, Mann does not gauge the probability of the action needed or how to promote it. Saying that, I continue to admire the updated accounts of how much or little previous planetary events provide probable prologues to those of the near future.

15. For one review of this history, and close engagements with Thomas Burnet, James Hutton, and Charles Lyell as key carriers of it, see Stephen Jay Gould, *Time's Arrow, Time's Cycle: Myth and Metaphor in the Discovery of Geological Time* (Cambridge, MA: Harvard University Press, 1987).

16. For a discussion that focuses more on Einstein, see William E. Connolly, *Resounding Events: Adventures of an Academic from the Working Class* (New York: Fordham University Press, 2022), 176–85.

17. Michel Serres, *Genesis,* trans. Geneviève James and James Nielson (Ann Arbor: University of Michigan Press, 1995), 5.

18. Michel Serres with Bruno Latour, *Conversations on Science, Culture, and Time,* trans. Roxanne Lapidus (Ann Arbor: University of Michigan Press, 1995), 105.

19. Michel Serres, *Branches: A Philosophy of Time, Event and Advent,* trans. Randolph Burks (London: Bloomsbury, 2020), 125.

20. Michel Serres, *The Incandescent,* trans. Randolph Burks (London: Bloomsbury, 2018), 103.

21. Serres with Latour, *Conversations*, 58.

22. Serres, *Incandescent*, 161. I should also note two other books that have helped me to come to terms with Serres. The first, *Michel Serres: Figures of Thought* (Edinburgh: Edinburgh University Press, 2020), by Christopher Watkin, explores the development of Serres's orientation to nature, logic, and time across several decades. The second, *Time and History in Deleuze and Serres*, ed. Bernd Herzogenrath (New York: Continuum International, 2012), helped me to compare Deleuze, a figure more familiar, with Serres. Jane Bennett and I have an essay in the latter volume entitled "The Crumpled Handkerchief," 153–72.

23. I should perhaps note here that chapter 4 of Connolly, *Facing the Planetary: Entangled Humanism and the Politics of Swarming* (Durham, NC: Duke University Press, 2017), advanced the thesis of "bumpy temporalities" periodically intersecting or colliding to turn the direction of time. I had not yet read the Michel Serres discussion of this topic. The thing now is to profit from that exposure and to extend the earlier account.

24. Perhaps this is the moment to note the thoughtful book by Stewart Motha, *Archiving Sovereignty: Law, History, Violence* (Ann Arbor: University of Michigan Press, 2018). Motha charts how gulag islands have served as Indian and Australian detention centers for refugees in the Indian and Pacific Oceans; they spawn a new imperialism that stops refugee flows and is horrendous in its application. As the analysis proceeds, he shows how "heterogeneous time" interrupts singular, progressive images that have prevailed in the west. By heterogeneous time, I think he means how impositions and aspirations from the past suddenly erupt in the present, turning current flows and making time less linear than generally projected. I profit from what Motha says, and hope his statements about Australian and Indian imperialism coalesce with those made in earlier chapters about settler societies in the Americas. Time as multiplicity, however, adds a couple of dimensions to his story of time. Nonhuman trajectories of different sorts also participate in time as multiplicity. So the image invoked here is both heterogeneous in Motha's sense and more multiple in the types of trajectories it includes.

25. I listed some of these events in a recent essay, "The Obduracy of the Event and the Tasks of the Intellectual," in *The Long 2020*, ed. Richard Grusin and Maureen Ryan (Minneapolis: University of Minnesota Press, 2022), 89–105. There the prime task was to define the tasks of the intellectual in a world in which manifold events interrupt an image of smooth, linear time. Now the intent is to support the theme of time as a multiplicity of divergent temporalities.

26. For a discussion of the import of the film *Melancholia* for human exceptionalism see William E. Connolly, *The Fragility of Things: Self-Organizing Processes, Neoliberal Fantasies, and Democratic Activism* (Durham, NC: Duke University Press, 2013), 43–52.

27. See Michael Hathaway, *What a Mushroom Lives For: Matsutake and the Worlds They Make* (Princeton, NJ: Princeton University Press, 2022), chap. 1.

28. Michael Benton, *When Life Nearly Died: The Greatest Mass Extinction of All Time* (London: Thames Hudson, 2005).

29. See Wally Broecker, *The Great Ocean Conveyor: Discovering the Trigger for Abrupt Climate Change* (Princeton, NJ: Princeton University Press, 2010).

30. For one geological account (that has also been contested), see William Ryan and Walter Pitman, *Noah's Flood: The New Scientific Discoveries about the Event That Changed History* (New York: Simon and Schuster, 1998).

31. See Mike Davis, *Late Victorian Holocausts: El Niño Famines and the Making of the Third World* (New York: Verso Books, 2001). Davis, at odds with most western social theory of the day, combined a history of the colonial holocaust with an account of how intense El Niños formed and altered the wind patterns over India. A Marxist social theorist who refused relatively early on to succumb to sociocentrism.

32. In chapter 4, Carlo Rovelli's perspectivism was placed into conversation with Amazonian cosmologies. It is feasible, too, to suggest that the quantum theorist's image of time possesses parallels with that of Michel Serres. Here we merely record a couple of suggestive quotes from Rovelli's *Order of Time* (New York: Riverhead Books, 2015). "It is not possible to think of duration as continuous. We must think of it as discontinuous, not as something that flows uniformly but as something that in a certain sense jumps, kangaroo-like, from one value to another" (84). "On closer inspection, in fact, even the things that are most 'thinglike' are nothing more than long events. The hardest stone . . . is in reality a complex vibration of quantum fields, . . . a process that for a brief moment manages to keep its shape" (98).

33. See Kohei Saito, *Karl Marx's Ecosocialism.* In chapter 6 Saito reviews Marx's late engagements with the soil scientists Leibig and Fraas. According to the author, these readings, plus the disappointments of the 1848 defeat, encouraged Marx to place ecology at the center of his critique of capitalism and also pulled him to explore other times when rifts between culture and nature were in part fashioned by patterns of agriculture that promoted desertification. Marx, of course, did not explore the larger planetary processes that must also be included in such accounts. But the engagements he did pursue were reflective and helped to force significant revisions in the thought of his middle phase. That is to say, a growing awareness of this shift in Marx helps to set the stage for reflective engagements between Marxists and others on the dynamics of the Anthropocene. The Saito book is important for both Marxists and non-Marxists.

34. For a robust review of this and other examples of creative activism see the discussion of "Blockadia" in Naomi Klein, *This Changes Everything: Capitalism vs. The Climate* (New York: Simon and Schuster, 2014). Klein, indeed, begins by demeaning the role experiments I include as one component in a politics of swarming. But later she implicitly includes them, as she explores how her pregnancy encouraged her to become much more alert to the precarity of emergent life today for multiple species and to think about how ranchers and urban ecologists had to work on themselves to forge new alliances with the Lakota.

35. Two recent books help us to ponder such connections. The first, Jane Bennett, *Influx and Efflux: Writing Up with Walt Whitman* (Durham, NC: Duke University Press, 2021), focuses on how cultivation of the "middle voice"—a voice dampened in European languages—can help us to receive experiences and deepen earth

attachments muted by the subject/object duality invested in the English language and other institutions. The second, David Hinton, *Wild Mind, Wild Earth: Our Place in the Sixth Extinction* (Boulder, CO: Shambhala, 2022), places Chinese Taoist perspectives into conjunction with Amerindian perspectives, deploying both to allow muted features of western life to be tapped and expanded. In each case, I think, the quest is not to replace projects of earth mastery with one of dwelling in harmony with an organic, smooth earth, but to become attached to the grandeur of elements of wildness that link us to the earth. Meditation, for instance, does not necessarily have to be attached to an organic image of the earth. This is an ecological insight Nietzsche, too, elaborated in his work. He saw how appreciation of an earth of bumpy emergences means attention to how it is often not oriented to human comfort or primacy. To love the earth, he contended, is to fold appreciation of its periodic volatility into that love, as you affirm some turns and struggle against others.

36. For one lovely text that explores how to weave role experiments into larger social fabrics see adrienne maree brown, *Emergent Strategy: Shaping Change, Changing Worlds* (Chico, CA: AK, 2017). To me, her text links the philosophy of emergence explored at various moments in this text to the politics of swarming. Here is one quote to suggest the gist of its explorations: "Transformation doesn't happen in a linear way, at least not one we can always track. It happens in cycles, convergences, explosions" (105).

Bibliography

Alchon, Suzanne Austin. *A Pest in the Land: New World Epidemics in a Global Perspective*. Albuquerque: University of New Mexico Press, 2003.

Andrist, Ralph. *The Long Death: The Last Days of the Plains Indians*. New York: Collier Books, 1964.

Anker, Elisabeth R. *Ugly Freedoms*. Durham, NC: Duke University Press, 2022.

Apess, William. *On Our Own Ground: The Complete Writings of William Apess, a Pequot*. Edited by Barry O'Connell. Amherst: University of Massachusetts Press, 1992.

Augustine. *The City of God Against the Pagans*. Translated by R. W. Dyson. Cambridge: Cambridge University Press, 1998.

———. *The Confessions of St. Augustine*. Translated by John Kenneth Ryan. New York: Image Books, 1960.

———. *Select Letters*. Translated by James Houston Baxter. Cambridge, MA: Harvard University Press, 1930.

Baldwin, James. *Another Country*. New York: Vintage Books, 1960.

———. *The Fire Next Time*. New York: Vintage Books, 1963.

———. *Notes of a Native Son*. Boston: Beacon, 1955.

Beaton, Roderick. *The Greeks: A Global History*. New York: Basic Books, 2021.

Belew, Kathleen. *Bring the War Home: The White Power Movement and Paramilitary America*. Cambridge, MA: Harvard University Press, 2018.

Bennett, Jane. *Influx and Efflux: Writing Up with Walt Whitman*. Durham, NC: Duke University Press, 2020.

———. *Vibrant Matter: A Political Ecology of Things*. Durham, NC: Duke University Press, 2010.

Bennett, Jane, and William E. Connolly. "The Crumpled Handkerchief." In *Time and History in Deleuze and Serres*, edited by Bernd Herzogenrath, 153–72. New York: Continuum International, 2012.

Benton, Michael. *When Life Nearly Died: The Greatest Mass Extinction of All Time.* London: Thames Hudson, 2005.
Berdan, Frances F. *The Aztecs: Lost Civilizations.* London: Reaktion Books, 2021.
Best, Michael H., and William E. Connolly. *The Politicized Economy.* Lexington, MA: D. C. Heath, 1976. Rev. ed., 1983. See esp. "Nature and Its Largest Parasite."
Blackhawk, Ned. *The Rediscovery of America: Native Peoples and the Unmaking of U.S. History.* New Haven, CT: Yale University Press, 2023.
Blom, Philipp. *Nature's Mutiny: How the Little Ice Age of the Long Seventeenth Century Transformed the West and Shaped the Present.* New York: Liverwright, 2019.
Bloom, Harold, ed. *The Book of J.* Translated by David Rosenberg and interpreted by Harold Bloom. New York: Grove Weidenfeld, 1990.
Blow, Charles. "The Politics of Persecution." *New York Times*, August 14, 2022.
Blumenberg, Hans. *The Legitimacy of the Modern Age.* Translated by Robert M. Wallace. Cambridge, MA: MIT Press, 1983.
Brendese, P. J. *Segregated Time.* New York: Oxford University Press, 2023.
Broecker, Wally. *The Great Ocean Conveyor: Discovering the Trigger for Abrupt Climate Change.* Princeton, NJ: Princeton University Press, 2010.
brown, adrienne maree. *Emergent Strategy: Shaping Change, Changing Worlds.* Chico, CA: AK, 2017.
Bruyneel, Kevin. *Settler Memory: The Disavowal of Indigeneity and the Politics of Race in the United States.* Chapel Hill: University of North Carolina Press, 2021.
Calvin, John. *The Institutes of Christian Religion.* Edited by Tony Lane and Hilary Osborne. Newark, NJ: Baker House, 1986.
Calvo, Paco, with Natalie Lawrence. *Planta Sapiens: The New Science of Plant Intelligence.* New York: W. W. Norton, 2022.
Canales, Jimena. *The Physicist and the Philosopher: Einstein, Bergson, and the Debate That Changed Our Understanding of Time.* Princeton, NJ: Princeton University Press, 2015.
Carter, Angela. *The Infernal Desire Machines of Doctor Hoffman.* London: Penguin Books, 1972.
Castro, Eduardo Viveiros de. *Cannibal Metaphysics.* Edited and translated by Peter Skafish. Minneapolis: University of Minnesota Press, 2014.
———. *The Inconstancy of the Indian Soul: The Encounter of Catholics and Cannibals in 16th-Century Brazil.* Translated by Gregory Duff Morton. Chicago: Prickly Paradigm, 2011.
———. *The Relative Native: Essays on Indigenous Conceptual Worlds.* Chicago: Hau Books, 2015.
Commoner, Barry. *The Closing Circle: Nature, Man, and Technology.* New York: Knopf, 1971.
Connolly, William E. *Aspirational Fascism: The Struggle for Multifaceted Democracy under Trumpism.* Minneapolis: University of Minnesota Press, 2017.
———. *The Augustinian Imperative: A Reflection on the Politics of Morality.* Newbury Park, CA: Sage, 1993.

———. "'Beyond Good and Evil': The Ethical Sensibility of Michel Foucault." *Political Theory* 21, no. 3 (August 1993): 365–89.
———. "Bodily Stresses, Cultural Drives, Fascist Contagions." *Theory & Event* 25, no. 3 (July 2022): 689–709.
———. *Capitalism and Christianity, American Style*. Durham, NC: Duke University Press, 2008.
———. *Climate Machines, Fascist Drives, and Truth*. Durham, NC: Duke University Press, 2019. See esp. chap. 3: "The Lure of Truth"
———. *The Ethos of Pluralization*. Minneapolis: University of Minnesota Press, 1995.
———. *Facing The Planetary: Entangled Humanism and the Politics of Swarming*. Durham, NC: Duke University Press, 2017.
———. *The Fragility of Things: Self-Organizing Processes, Neoliberal Fantasies, and Democratic Activism*. Durham, NC: Duke University Press, 2013.
———. *Identity/Difference: Democratic Negotiations of Political Paradox*. Ithaca, NY: Cornell University Press, 1991.
———. "The Obduracy of the Event and the Tasks of the Intellectual." In *The Long 2020*, edited by Richard Grusin and Maureen Ryan, 89–105. Minneapolis: University of Minnesota Press, 2022).
———. *Politics and Ambiguity*. Madison: University of Wisconsin Press, 1987.
———. "Race and the Anthropocene: Planetary Circuits of Imperial Power." *The Contemporary Condition* (blog), December 14, 2021. http://contemporarycondition.blogspot.com/2021/12/race-and-anthropocene-planetary.html.
———. *Resounding Events: Adventures of an Academic from the Working Class*. New York: Fordham University Press, 2022.
———. "Tocqueville, Territory, and Violence." *Theory, Culture & Society* 11, no. 1 (February 1994): 19–42.
———. *A World of Becoming*. Durham, NC: Duke University Press, 2011.
Connolly, William E., and Jairus Grove. "Planetary Events, Climate Catastrophes, and the Limits of the Human Sciences." In *Between Catastrophe and Revolution: Essays in Honor of Mike Davis*, edited by Daniel Monk and Michael Sorkin, 29–54. New York: OR Books, 2022.
Daly, Mary. *Beyond God the Father: Toward a Philosophy of Women's Liberation*. Boston: Beacon, 1973.
———. *The Mary Daly Reader*. Edited by Jennifer Rycenga and Linda Barufaldi. New York: New York University Press, 2017.
———. "The Metapatriarchal Journey of Exorcism and Ecstasy." In *The Mary Daly Reader*, edited by Jennifer Rycenga and Linda Barufaldi, 131–62. New York: New York University Press, 2017.
Damasio, Antonio R. *Descartes' Error: Emotion, Reason, and the Human Brain*. New York: Avon Books, 1994.
Davis, Mike. *Late Victorian Holocausts: El Niño Famines and the Making of the Third World*. New York: Verso Books, 2001.
———. *Old Gods, New Enigmas: Marx's Lost Theory*. New York: Verso Books, 2016.

Deacon, Terrence W. *Incomplete Nature: How Mind Emerged from Matter.* New York: W. W. Norton, 2012.

Deleuze, Gilles. *Difference and Repetition.* Translated by Paul Patton. Minneapolis: University of Minnesota Press, 1994.

Descartes, René. *Discourse on Method and Meditations.* Translated by Laurence J. Lafleur. Saddle River, NJ: Prentice Hall, 1952.

Dobyns, Henry F. "Estimating Aboriginal American Population: An Appraisal and Techniques with a New Hemispheric Estimate." *Current Anthropology* 77, no. 3 (October 1966): 395–449.

Dunbar-Ortiz, Roxanne. *An Indigenous Peoples' History of the United States.* Boston: Beacon, 2014.

——. *Not "A Nation of Immigrants": Settler Colonialism, White Supremacy, and a History of Erasure and Exclusion.* Boston: Beacon, 2019.

Flavelle, Christopher. "As Groundwater Dwindles, Powerful Players Block Change." *New York Times,* November 24, 2023. https://www.nytimes.com/interactive/2023/11/24/climate/groundwater-levels.html.

Foucault, Michel. "On the Genealogy of Ethics: Overview of Work in Progress." In *The Foucault Reader,* edited by Paul Rabinow, 340–72. New York: Pantheon Books, 1984.

Fox, Douglas. "The Coming Collapse." *Scientific American* 327, no. 5 (November 2022): 32–41.

Francis. *Laudato Si': On Care for Our Common Home.* Huntington, IN: Our Sunday Visitor, 2015.

Frankopan, Peter. *The Earth Transformed: An Untold Story.* New York: Alfred A. Knopf, 2023.

Ghosh, Amitav. *The Nutmeg's Curse: Parables for a Planet in Crisis.* Chicago: University of Chicago Press, 2021.

Gimbutas, Marija. *The Living Goddesses.* Edited by Miriam R. Dexter. Berkeley: University of California Press, 2001.

Ginzburg, Carlo. *The Cheese and the Worms: The Cosmos of a Sixteenth-Century Miller.* Translated by John Tedeschi and Anne C. Tedeschi. London: Routledge and Kegan Paul, 1980.

Glissant, Édouard. *Poetics of Relation.* Translated by Betsy Wing. Ann Arbor: University of Michigan Press, 2010.

Gould, Stephen Jay. *Time's Arrow, Time's Cycle: Myth and Metaphor in the Discovery of Geological Time.* Cambridge, MA: Harvard University Press, 1987.

Grandin, Greg. *The End of the Myth: From the Frontier to the Border Wall in the Mind of America.* New York: Henry Holt, 2019.

Grove, Jairus Victor. *Savage Ecology: War and Geopolitics at the End of the World.* Durham, NC: Duke University Press, 2019.

Habermas, Jürgen. *Legitimation Crisis.* Translated by Thomas McCarthy. Boston: Beacon, 1973.

Hales, Steven D. "Nietzsche on Logic." *Philosophy and Phenomenological Research* 56, no. 4 (December 1996): 819–35.

Hämäläinen, Pekka. *Indigenous Continent: The Epic Contest for North America.* New York: W. W. Norton, 2022.
Hardt, Michael, ed. *Thomas Jefferson: The Declaration of Independence.* New York: Verso, 2007.
Harper, Kyle. *The Fate of Rome: Climate, Disease, and the End of an Empire.* Princeton, NJ: Princeton University Press, 2017.
Hathaway, Michael J. *What a Mushroom Lives For: Matsutake and the Worlds They Make.* Princeton, NJ: Princeton University Press, 2022
Heller-Roazen, Daniel. *The Fifth Hammer: Pythagoras and the Disharmony of the World.* New York: Zone Books, 2011.
Herzogenrath, Bernd, ed. *Time and History in Deleuze and Serres.* New York: Continuum International, 2012.
Hesiod. *Theogony.* Translated and introduced by Norman O. Brown. Saddle River, NJ: Prentice Hall, 1953.
———. *Works and Days.* In *Hesiod: Theogony, Works and Days, Shield.* Translated and edited by Apostolos Athanassakis, 59–111. Baltimore, MD: Johns Hopkins University Press, 2004.
Hinton, David. *Wild Mind, Wild Earth: Our Place in the Sixth Extinction.* Boulder, CO: Shambhala, 2022.
Hirsch, Fred. *Social Limits to Growth.* London: Blackwell, 1977.
Honig, Bonnie. *Antigone, Interrupted.* Cambridge: Cambridge University Press, 2013.
James, William. *A Pluralistic Universe.* Lincoln: University of Nebraska Press, 1996.
Kant, Immanuel. *Anthropology from a Pragmatic Point of View.* Translated by Victor Dowdell. Carbondale: Southern Illinois University Press, 1978.
———. *Critique of Practical Reason.* Translated by Lewis White Beck. New York: Library of Liberal Arts, 1993.
———. "Of the Different Human Races." In *The Idea of Race,* edited by Robert Bernasconi and Tommy Lee Lott. Indianapolis: Hackett, 2000.
———. "On the Proverb: That May Be True in Theory but Is of No Practical Use." In *Perpetual Peace and Other Essay on Politics, History, and Morals,* translated by Ted Humphrey. Indianapolis: Hackett, 1983.
———. "To Perpetual Peace: A Philosophical Sketch." In *Perpetual Peace and Other Essays on Politics, History, and Morals,* translated by Ted Humphrey. Indianapolis: Hackett, 1983.
Keller, Catherine. *The Face of the Deep: A Theology of Becoming.* New York: Routledge, 2003.
———. *Facing Apocalypse: Climate, Democracy, and Other Last Chances.* Maryknoll, NY: Orbis Books, 2021.
———. *On the Mystery: Discerning Divinity in Process.* Minneapolis, MN: Fortress, 2008.
———. *Political Theology of the Earth: Our Planetary Emergency and the Struggle for a New Public.* New York: Columbia University Press, 2018.

Kennedy, Jonathan. *Pathogenesis: A History of the World in Eight Plagues.* New York: Crown, 2023.

Khan, Naveeda. *River Life and the Upspring of Nature.* Durham, NC: Duke University Press, 2023.

Klein, Naomi. *Doppelganger: A Trip into the Mirror World.* London: Penguin, 2023.

——. *This Changes Everything: Capitalism vs. The Climate.* New York: Simon and Schuster, 2014.

Kohn, Eduardo. "Anthropology of Ontologies." *Annual Review of Anthropology* 44 (August 2015): 311–27.

Krause, Sharon R. *Eco-emancipation: An Earthly Politics of Freedom.* Princeton, NJ: Princeton University Press, 2023.

Lamb, Hubert. *Climate: Present, Past and Future.* Vol. 2, *Climatic History and the Future.* New York: Routledge, 1977.

Larsen, Svend. "The Lisbon Earthquake and the Scientific Turn in Kant's Philosophy." *European Review* 14, no. 3 (July 2006): 359–67.

Lawtoo, Nidesh. *The Phantom of the Ego: Modernism and the Mimetic Unconscious.* East Lansing: Michigan State University Press, 2013.

Lipsky, David. *The Parrot and the Igloo: Climate and the Science of Denial.* New York: W. W. Norton, 2023.

Lucanus, Ocellus. "On the Nature of the Universe." In *The Pythagorean Sourcebook and Library,* compiled and translated by Kenneth Guthrie. Grand Rapids, MI: Phanes, 1987.

MacLean, Nancy. *Democracy in Chains: The Deep History of the Radical Right's Stealth Plans for America.* New York: Penguin Books, 2017.

Maffie, James. *Aztec Philosophy: Understanding a World in Motion.* Boulder: University Press of Colorado, 2014.

Mann, Michael E. *Our Fragile Moment: How Lessons from Earth's Past Can Help Us Survive the Climate Crisis.* New York: PublicAffairs, 2023.

Manschot, Henk. *Nietzsche and the Earth: Biography, Ecology, Politics.* Translated by Liz Waters. London: Bloomsbury, 2021.

Martindale, Charles. Introduction to *Ovid Renewed: Ovidian Influences on Literature and Art from the Middle Ages to the Twentieth Century,* edited by Charles Martindale. Cambridge: Cambridge University Press, 1988.

McGuire, Bill. *Hothouse Earth: An Inhabitant's Guide.* London: Icon Books, 2022.

——. *Waking the Giant: How a Changing Climate Triggers Earthquakes, Tsunamis and Volcanoes.* Oxford: Oxford University Press, 2012.

McLarin, Kim. *James Baldwin's* Another Country: *Bookmarked.* New York: Ig, 2021.

Mills, Charles W. *Black Rights/White Wrongs: The Critique of Racial Liberalism.* New York: Oxford University Press, 2017.

Mitchell, Stephen, ed. *The Book of Job.* San Francisco: Northpoint, 1979.

Montgomery, David R. *The Rocks Don't Lie: A Geologist Investigates Noah's Flood* . New York: W. W. Norton, 2012.

Motha, Stewart. *Archiving Sovereignty: Law, History, Violence.* Ann Arbor: University of Michigan Press, 2018.

Nietzsche, Friedrich. *Thus Spoke Zarathustra: A Book for None and All.* Translated by Walter Kaufmann. New York: Penguin Books, 1968.

———. *Twilight of the Idols and The Anti-Christ.* Translated by R. J. Hollingdale. New York: Penguin, 1968.

———. *The Will to Power.* Translated by Walter Kaufmann and R. J. Hollingdale. New York: Vintage Books, 1967.

———. *Writings from the Late Notebooks.* Edited by Rüdiger Bittner. Translated by Kate Sturge. Cambridge: Cambridge University Press, 2003.

Nixey, Catherine. *The Darkening Age: The Christian Destruction of the Classical World.* New York: Macmillan, 2017.

Olivier, Guilhem. *Mockeries and Metamorphoses of an Aztec God: Tezcatlipoca, "Lord of the Smoking Mirror."* Translated by Michel Besson. Boulder: University Press of Colorado, 1997.

Olson, Joel. *The Abolition of White Democracy.* Minneapolis: University of Minnesota Press, 2004.

Ovid. *Metamorphoses.* Translated and edited by Charles Martin. New York: W. W. Norton, 2004.

Plumwood, Val. *The Eye of the Crocodile.* Edited by Lorraine Shannon. Canberra: Australian National University E Press, 2012.

Quine, W. V., and J. S. Ullian. *The Web of Belief.* 2nd ed. New York: Random House, 1978.

Radin, Paul. *The Trickster: A Study in American Indian Mythology.* New York: Schoeken Books, 1967.

Reinhardt, O., and D. R. Oldroyd. "Kant's Theory of Earthquakes and Volcanic Action." *Annals of Science* 40, no. 3 (July 1983): 247–72.

Restall, Matthew. *Seven Myths of the Spanish Conquest.* Rev. ed. Oxford: Oxford University Press, 2021.

Ricœur, Paul. *The Symbolism of Evil.* Translated by Emerson Buchanan. Boston: Beacon, 1967.

Romilly, Jacqueline de. *Time in Greek Tragedy.* Ithaca, NY: Cornell University Press, 1968.

Rovelli, Carlo. *Anaximander and the Birth of Science.* Translated by Marion Lignana Rosenberg. New York: Riverhead Books, 2021.

———. *Helgoland: Making Sense of the Quantum Revolution.* Translated by Erica Segre and Simon Carnell. New York: Riverhead Books, 2020.

———. *The Order of Time.* Translated by Erica Segre and Simon Carnell. New York: Riverhead Books, 2015.

Russell, Bertrand. *The ABC of Relativity.* New York: Harper and Brothers, 1925.

Ryan, William, and Walter Pitman. *Noah's Flood: The New Scientific Discoveries about the Event That Changed History.* New York: Simon and Schuster, 1998.

Sahagún, Bernardino de. *A History of Ancient Mexico.* Translated by Fanny R. Bandelier. New York: Fisk University Press, 1932.

Saito, Kohei. *Karl Marx's Ecosocialism: Capital, Nature, and the Unfinished Critique of Political Economy.* New York: Monthly Review Press, 2017.

Segall, Matthew David. *Crossing the Threshold: Etheric Imagination in the Post-Kantian Process Philosophy of Schelling and Whitehead.* Olympia: Integral Imprint, 2023.

Serres, Michel. *The Birth of Physics.* Translated by Jack Hawkes. New York: Clinamen, 2001.

——. *Branches: A Philosophy of Time, Event and Advent.* Translated by Randolph Burks. London: Bloomsbury, 2020.

——. *Genesis.* Translated by Geneviève James and James Nielson. Ann Arbor: University of Michigan Press, 1995.

——. *The Incandescent.* Translated by Randolph Burks. London: Bloomsbury, 2018.

Serres, Michel, with Bruno Latour. *Conversations on Science, Culture, and Time.* Translated by Roxanne Lapidus. Ann Arbor: University of Michigan Press, 1995.

Simmel, Georg. *The Sociology of Georg Simmel.* Edited by Kurt Wolff. New York: Free Press, 1962.

Sophocles. *Oedipus the King.* In *The Oedipus Plays of Sophocles*, translated by Paul Roche. New York: Penguin, 1996.

——. *Oedipus the King.* In *Sophocles*, vol. 2 of *The Complete Greek Tragedies*, translated by David Grene. Edited by David Grene and Richmond Lattimore. Chicago: University of Chicago Press, 1992.

Strauss Clay, Jenny. *Hesiod's Cosmos.* Cambridge: Cambridge University Press, 2003.

Taylor, Charles. "Interpretation and the Sciences of Man." *Review of Metaphysics* 25, no. 1 (Fall 1971): 3–51.

——. *A Secular Age.* Cambridge, MA: Belknap, 2007.

Thucydides. *The History of the Peloponnesian War.* Translated by Richard Crawley. New York: Compass, 2019.

Tocqueville, Alexis de. *Democracy in America.* Translated by George Lawrence. New York: Anchor Books, 1969.

Todorov, Tzvetan. *The Conquest of America: The Question of the Other.* Translated by Richard Howard. New York: Harper and Row, 1984.

——. *In Defence of the Enlightenment.* Translated by Gila Walker. London: Atlantic Books, 2006.

——. *On Human Diversity: Nationalism, Racism, and Exoticism in French Thought.* Translated by Catherine Porter. Cambridge, MA: Harvard University Press, 1993.

Townsend, Camilla. *Fifth Sun: A New History of the Aztecs.* Oxford: Oxford University Press, 2019.

Tripaldi, Laura. *Parallel Minds: Discovering the Intelligence of Materials.* Cambridge, MA: MIT Press, 2021.

Trouillot, Michel-Rolph. *Silencing the Past: Power and the Production of History.* Boston: Beacon, 1995.

Turetzky, Philip. *Time.* New York: Routledge, 1968.

Vermes, Geza. *The Religion of Jesus the Jew.* Minneapolis, MN: Fortress, 1993.

Vrdoljak, Tvrtko. "The Metaphysics of Politics: Nietzsche and His Interlocutors." PhD dissertation, Johns Hopkins University, 2023.

Watkin, Christopher. *Michel Serres: Figures of Thought*. Edinburgh: Edinburgh University Press, 2020.

Weber, Max. *The Protestant Ethic and the Spirit of Capitalism*. Translated by Talcott Parsons. New York: Charles Scribner's Sons, 1958.

Welz, Adam. *The End of Eden: Wild Nature in the Age of Climate Breakdown*. New York: Bloomsbury, 2023.

West, Cornel. *Black Prophetic Fire*. Edited by Christa Buschendorf. Boston: Beacon, 2014.

White, Sam. A *Cold Welcome: The Little Ice Age and Europe's Encounter with North America*. Cambridge, MA: Harvard University Press, 2017.

Whitemarsh, Tim. *Battling the Gods: Atheism in the Ancient World*. New York: Alfred A. Knopf, 2015.

Williams, Bernard. *Shame and Necessity*. Berkeley: University of California Press, 1993.f

Index

WILLIAM E. CONNOLLY is Krieger-Eisenhower Professor emeritus at Johns Hopkins, where he teaches political theory. His books include *Resounding Events* (Fordham, 2022); *Climate Machines, Fascist Drives, and Truth* (Duke, 2020); *Aspirational Fascism* (Minnesota, 2017); *Facing the Planetary* (Duke, 2017); *Capitalism and Christianity, American Style* (Duke, 2008); *Why I Am Not a Secularist* (Minnesota, 1999); *The Ethos of Pluralization* (Minnesota, 1995); and *The Terms of Political Discourse* (Princeton, 1983; 3rd ed., 1993). In a poll of American political theorists published in 2010, he was named the fourth most influential political theorist in America over the last twenty years, after Rawls, Habermas, and Foucault.

www.ingramcontent.com/pod-product-compliance
Lightning Source LLC
Jackson TN
JSHW081132140125
77106JS00002B/35

9781531509217